Dominance and Reproduction in Baboons *(Papio cynocephalus)*
A Quantitative Analysis

Contributions to Primatology

Vol. 7

Editors

H. KUHN, Göttingen
W. P. LUCKETT, New York, N.Y.
C. R. NOBACK, New York, N.Y.

A. H. SCHULTZ, Zürich
D. STARCK, Frankfurt a.M.
F. S. SZALAY, New York, N.Y.

S. Karger · Basel · München · Paris · London · New York · Sydney

Dominance and Reproduction in Baboons (Papio cynocephalus)

A Quantitative Analysis

GLENN HAUSFATER

Department of Psychology, University of Virginia, Charlottesville, Va.

24 figures and 37 tables, 1975

S. Karger · Basel · München · Paris · London · New York · Sydney

Contributions to Primatology

(Successor to 'Bibliotheca Primatologica')

S. Karger · Basel · München · Paris · London · New York · Sydney
Arnold-Böcklin-Strasse 25, CH–4011 Basel (Switzerland)

Contents

Contents

IV. The Mating System of Amboseli Baboons

V. Estrous Females and Group Organization

VI. Conclusions and Speculations

'However, what emerges... is that quantitative analysis of communication has the seeds of detailed explanation in the future but so far tells us less than does intelligent empathy' [JOLLY, 1972b, p. 149].

Acknowledgments

This research was made possible only through the hard work, patience, tolerance, support, and cooperation of friends and loved ones. A few individuals deserve special mention.

STUART and JEANNE ALTMANN provided clear standards of excellence in thought and research on animal behavior against which I measured my own work at every stage in the data collection, analysis and writing.

SUE ANN HAUSFATER added moments of joy to my life in Amboseli and in Chicago.

MWONGI KIREGA and SUSAN ZIMMER provided friendship and conscientious service without which this research would have never been completed.

In addition, I gratefully acknowledge permission to carry out research in Kenya from the Office of the President, Game Department, and National Parks of Kenya. I also gratefully acknowledge permission to carry out research and cooperation from the Olkejuado County Council and the Amboseli game wardens. This research was supported by NSF grant GB-27170, NIMH grant MH 19,617, NIMH predoctoral fellowship 5 FO1 MH 47385, and NIMH Mental Health small grant MH 24473-01, 1 RO3 MSM. The generous support of the above agencies is appreciated.

I. Introduction

In August, 1971, I began a 14-month study of the relationship between agonistic dominance and reproductive success in a group of free-living yellow baboons *(Papio cynocephalus)*. The purpose of this research was to determine whether the dominance rank of male baboons, an aspect of their individual social behavior, was related to their reproductive success. Are differences in dominance among males accompanied by differential reproduction? Does the dominant male of a baboon social group have a higher probability of mating with fertile females, and thus of leaving offspring, than do lower ranking males? Specifically, I attempted to determine how the dominance rank of male baboons was related to their probability of mating with a sexually cycling female on, or near, her day of ovulation. Such differential mating can result in differential reproduction within one sex which is the essence of sexual selection.

Yellow baboons *(Papio cynocephalus)*, the subjects of the present study, live in closed social groups composed of several adult males, several adult females, and the females' offspring. No permanent male-female pair bonds are formed in these groups and thus each male in the group is potentially in competition with all other males for access to receptive females. Of course, females are able to mate preferentially with one or more males. This description is also applicable to the two other savannah baboon species, olive baboons *(P. anubis)* [DeVore, 1965; Rowell, 1966a] and chacma baboons *(P. ursinus)* [Saayman, 1970; Stoltz and Saayman, 1970].

In contrast, the basic social group of hamadryas baboons *(P. hamadryas)* is a one-male unit composed of one adult male, several adult females and their offspring and, occasionally, a subadult male [Kummer, 1968]. All mature females live in one-male units, while some 20% of adult males live outside of these units. Adult males are believed to mate exclusively with females in their own units; thus, the reproductive success of an adult male compared to other adult males in the same population depends primarily on the number of mature females in his unit and the length of time he controls the unit.

Both the polygamous mating system of hamadryas baboons and the competitive-preferential mating system of savannah baboons may result in differential reproduction among adult males. Insofar as any male- or female-specific physiological, morphological or behavioral trait in baboons has a genetic component, differential reproduction by the individuals exhibiting this trait will result in the evolution and maintenance of sexual dimorphism and diethism. In fact, all species of the genus *Papio* exhibit marked sexual dimorphism. Adult males weigh approximately twice as much as adult females and the two sexes show numerous other differences in appearance and behavior [FREEDMAN, 1963; SNOW and VICE, 1965]. A brief review of the theory of sexual selection follows below.

Sexual Selection

DARWIN [1871] wrote that sexual selection depended on the success of certain individuals over others of the same sex in relation to the propagation of their species, while natural selection depended on the success of both sexes at all ages in relation to the general conditions of life. The unsuccessful competitor in sexual selection did not suffer death necessarily; he merely produced fewer offspring.

Sexual selection was proposed by DARWIN to account for the secondary sexual characteristics of animals, particularly distinctive coloration [cf. HUXLEY, 1938], horns and antlers. Sexual selection also was considered to account for sexual dimorphism and diethism within a single species. It is interesting to note that no animal adornments 'perplexed' DARWIN so much as the 'brightly coloured hinder ends and adjoining parts of certain monkeys' [DARWIN, 1876, p. 18].

DARWIN [1871] believed that sexual selection operated through the action of two basic behavioral mechanisms: male competition for females, and female preference for certain males. In contrast, WALLACE [1889] rejected the notion of female preference as a means of sexual selection, but accepted the importance of male competition for mates. Later, FISHER [1930] pointed out that when a selective advantage is linked to a secondary sexual characteristic in one sex, there will be concurrent selection on the opposite sex for those individuals who most clearly discriminate this characteristic and thus most prefer the advantageous type. Therefore, male competition and female preference should be thought of as complementary, rather than alternative, mechanisms of sexual selection, and either one or both of these mechanisms may be operating to produce differential reproduction within a population of animals.

Differential Reproduction in Vertebrates

Differential reproduction by individuals of certain genotypes, temperaments, physique, behavior, or social status can have important consequences for the life history of the individuals involved [TRIVERS, 1972; ORIANS, 1969], for the organization of their social groups [GOSS-CUSTARD et al., 1972], and for the future genetic composition of their populations [WRIGHT, 1921; FISHER, 1930, CUSHING, 1941; LEWONTIN et al., 1968; ECKLAND, 1972].

It is becoming increasingly clear that among vertebrate species differential reproduction among individuals is the rule rather than the exception. Differential reproduction or mating success has been reported in fishes [NOBLE, 1938; BRAWN, 1961; GANDOLFI, 1971], reptiles and amphibians [NOBLE and BRADLEY, 1933; EVANS, 1938], birds [SKARD, 1937; SCOTT, 1942; WILEY, 1973], and mammals [BARTHOLOMEW, 1952; LEVINE et al., 1965; LEBOEUF, 1972]. Differential reproduction in invertebrates also has been reported [PARKER, 1970].

Within the order Primates, differential reproduction or mating success based on dominance rank [DEVORE, 1965; CARPENTER, 1942a], age [KUMMER, 1968; LOY, 1971] and geological factors [SADE, 1968] has been described. In particular, differential fertility and reproduction in man has been implicated in several chronic diseases [NEEL et al., 1965].

Baboons share in common with all other mammals the problem of selecting a mate. The physiology and behavior of mammalian reproduction is such that a female can be fertilized by only one male during each period of fertility (for exceptions see BEATTY, [1960] and BIRDSALL and NASH [1973]). Thus, female mammals usually cannot increase the number of offspring they produce by mating with more than one male per fertile period, but male mammals can increase their reproductive success by mating sequentially with more than one fertile female. It follows that the cost of an error in mate selection is more serious for a female than for a male [ORIANS, 1969]. Since potential mates are sufficiently plentiful that rejection of one mate usually has little consequence compared to the advantage of later selecting a more superior mate, it is to be expected that selectivity of mating by females will be practiced whenever possible and that mate preferences by females will be strongly exhibited [FISHER, 1930; ORIANS, 1969].

Study Subjects and Methods

Study Site

This study was conducted in the Masai-Amboseli Game Reserve, Kenya. Amboseli is an area of semi-arid short-grass savannah with a scattered cover of *Acacia tortilis* trees (fig. 1). Near the eastern edge of the Reserve lies the Enkongo Narok and Loginye swamps and an adjacent series of permanent water holes. The swamps and water holes are fed by subterranean run-off from Mt. Kilimanjaro and provide a year-round source of water for baboons and numerous other animal species.

Near the water holes and swamps, *Acacia xanthophloea*, the yellow barked acacia or fever tree, is the predominant tree species and occurs in association with the shrubs *Azima tetracantha* and *Salvadora persica* (fig. 1). In recent years the fever trees and associated vegetation have suffered high mortality as a result of a rise in the water table, increased soil salinity, and other factors [WESTERN and VAN PRAET, 1973].

Figure 2 shows the mean daily maximum and minimum temperatures by month and the total monthly rainfall at our camp, about ten miles east of the study area, but in the same general habitat. The long-term rainfall records for the Amboseli basin show that the annual rainfall is usually restricted to two periods of the year and totals 10–20 in. The 'short rains' usually fall in November and December, while the 'long rains' fall in March through April or May [GRIFFITHS, 1969].

Within the reserve, the study area itself was bounded roughly on the northeast by the Enkongo Narok swamp, on the northwest by Lake Amboseli, and on the south by a line joining the peaks of Endoinyo Nairabala, Endoinyo Ositet, and Ilmeireshari, three prominent hills in the area (Amboseli map, Survey of Kenya, Series Y731, Sheet 181/1, Ed. 2-SK). This is the same study area used by ALTMANN and ALTMANN [1970] in 1963–64 and subsequent years. Masai tribesmen lived and grazed their cattle within the study area.

The Baboon Population

The ecology of Amboseli baboons has been described by ALTMANN and ALTMANN [1970] who reported a baboon population of 2,600 animals in the study area in 1963–64. In August, 1971, when this study was begun, the baboon population in the same area was just over 200 animals, organized into seven social groups. One of these groups, Kitirua South, was observed in the study area less than ten times, though the other six groups were frequently

A

B

Fig. 1. The Amboseli study area. *A* Study group in short-grass savannah away from permanent water sources. Second tree from left is *Acacia tortilis*, the umbrella tree. *B* Study group in woodlands surrounding permanent water holes. Ground cover is a dense mat of the grass *Cynodon dactylon*, shrubs and bushes are *Azima tetracantha* and *Salvadora persica*, and fallen and standing trees are *Acacia xanthophloea*, the fever tree.

encountered. The cause of the marked decline in the Amboseli baboon population between 1963 and 1971 is unknown, but presumably is related to the deterioration of fever trees and associated vegetation.

Table I gives the size and composition of each group in the study area on one date near the beginning and one date near the end of the study period. The census categories used in table I and elsewhere in this work follow ALTMANN and ALTMANN [1970] and are summarized briefly as follows:

The infant-1 (I 1) and infant-2 (I 2) classes include individuals of both sexes from birth to six months of age and from six months to one year of age, respectively. The juvenile-1 (J 1) class includes animals of both sexes who are one and two years of age. The juvenile-2 (J 2) class is also composed of animals of both sexes and includes individuals of three and four years of age. Females are graduated from the J 2 class directly into the adult (Ad) class at about four years of age when they begin regular sexual cycles. Males are graduated from the J 2 class to the subadult (Sub) class when between four and six years of age and into the Ad class when older than six years of age.

The Study Group

Systematic observations on 'Alto's group', one of the seven groups in the study area, was begun on 1 August 1971 and continued almost daily through 3 September 1972. Alto's group was chosen from among the seven because it was small enough to allow recognition of every individual in the group, but large enough to provide a moderate sample of cycling females. The name and age-sex class of each individual in the group at the beginning of the study is given in table II, while table III lists all changes in the composition of the group that occurred during the study.

Each of the animals in the study group was individually identifiable by face, physique, and tail shape. An attempt was made to record the individual identity of the participants in every observed social interaction. Observations were made daily from approximately 0800 to 1800 h as the study group was followed in a Toyota Landcruiser. Field notes were dictated into a portable tape recorder and later transcribed verbatim.

Behavior Sampling Techniques

Information about baboon behavior was collected through several different sampling techniques; the choice of a particular behavior sampling

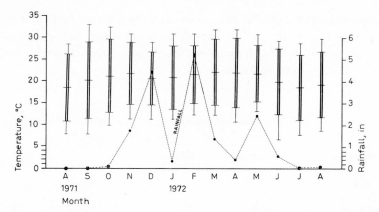

Fig. 2. Temperature and rainfall in the study area. The figure shows the daily mean minimum and mean maximum temperature (°C) by month and monthly rainfall (in inches) for the Amboseli study area during 1971–72. Double bars show daily mean minimum and mean maximum temperatures; single bars show monthly temperature extremes or range.

Table I. Size and composition of all baboon groups in the Amboseli study population. Two censuses are given for each group, one census from a date near the beginning of the study and one census from a date near the end of the study

Group name	Census date	Males						Females					Sex unknown					Total	+	− Error Factor[1]
		Ad	Sub	J2	J1	I2	I1	Ad	J2	J1	I2	I1	Ad	J2	J1	I2	I1			
Limp's group	14 Sept. 71	2			1			6								2	4	15	0	0
	6 Aug. 72	1	2	3				5	1									12	1	0
Stud's group	11 Sept. 71	14	5	3	6			23	1	3					1	4	8	68	0	0
	25 July 72	21	1	7	7			19	4	3					1	1	1	65	1	0
Hook's group	13 Sept. 71	11	1	1	4			13	2	1	1					1	3	38	0	1
	4 Aug. 72	6	3	2				13	1	1	1		2				3	32	2	0
BTF group	26 Aug. 71	2						5					1			3	2	13	4	0
	25 July 72	1		1	2	1		4	1	1								11	0	0
High Tail's group	26 Aug. 71	1						5					1	2	4		1	14	0	0
	24 July 72	1		1				2	4			1						9	0	0
Alto's group	1 Aug. 71	8		2	4	4	2	13	1	2								36	0	0
	31 July 72	8	2	4	2	2	1	10	1	2	2	1						35	0	0
Kitirua South group	6 Sept. 71																	30	0	0
	12 April 72	6	1	2	1			12		1					2	2	5	32	0	0

1 Follows Altmann and Altmann [1970].

technique depended on the research question being asked and the importance of that question in terms of the overall research project. In the discussion below, ALTMANN's [1974] terminology and classification of observational sampling methods will be followed.

Table II. Age, sex, and individual composition of Alto's group, 1 August 1971

Males	Females
Adults	
BJ	Alto
Cowlick	Fluff
Dutch	Judy
Ivan	Lulu
Max	Mom
Peter	New
Sinister	Oval
Stubby	Preg
	Ring
	Scar
	Skinny
	TT
	Twisty
Subadults	
—	—
Juvenile-2	
Ben	
Even	
Juvenile-1	
Red	Fem
Russ	
Stiff	
Stu	
Infant-2	
Bob	Gin
Major	Vee
Spot	
Swat	
Infant-1	
Herm	
Kub	

Table III. Changes in composition of Alto's group, 1 August 1971 through 3 September 1972

Date	Age-sex class	Individual	Event	Gain/loss	New group size
8/5/71	Ad♀	New	death	−1	35
8/5/71	Ad♂	BJ	emigration	−1	34
8/6/71	I1♂	Dogo	birth (mother: Ad♀ Preg)	+1	35
8/27/71	Ad♀	Twisty	death	−1	34
8/27/71	I1♂	Herm	death	−1	33
9/3/71	I2♂	Bob	death	−1	32
9/26/71	Ad♀	Skinny	death	−1	31
9/26/71	I2♂	Major	death	−1	30
10/28/71	Ad♂	BJ	immigration	+1	31
10/30/71	Ad♂	Cowlick	emigration	−1	30
10/31/71	Ad♀	Ring	emigration	−1	29
11/1/71	Ad♂	Crest	immigration	+1	30
11/1/71	Ad♂	Ivan	emigration	−1	29
11/3/71	Ad♂	Ivan	immigration	+1	30
11/3/71	Ad♀	Ring	immigration	+1	31
11/3/71	Ad♂	Cowlick	immigration	+1	32
11/7/71	Ad♂	Cowlick	emigration	−1	31
11/8/71	Ad♂	Cowlick	immigration	+1	32
11/22/71	Ad♂	Crest	emigration	−1	31
12/27–30/71	Ad♂	Sinister	emigration	−1	30
12/27–30/71	Ad♂	Peter	emigration	−1	29
1/1/72	J2♂	Ben	maturation to Sub	0	29
1/1/72	I2♂	Swat	maturation to J1	0	29
1/1/72	I1♂	Kub	maturation to I2	0	29
1/1/72	I1♂	Dogo	maturation to I2	0	29
1/2/72	Ad♂	Peter	immigration	+1	30
1/3/72	Ad♂	Sinister	immigration	+1	31
1/8/72	Ad♂	Max	emigration	−1	30
1/9/72	Ad♂	Max	immigration	+1	31
1/20/72	I1♀	Bell	birth (mother: Ad♀ TT)	+1	32
2/1/72	I1♀	Mindi	birth (mother: Ad♀ Scar)	+1	33
2/1/72	J1♂	Stiff	maturation to J2	0	33
2/1/72	J1♂	Stu	maturation to J2	0	33
2/1/72	J1♂	Red	maturation to J2	0	33
2/1/72	J1♂	Russ	maturation to J2	0	33
2/1/72	J1♀	Fem	maturation to J2	0	33
2/1/72	I2♂	Spot	maturation to J1	0	33
2/1/72	I2♀	Vee	maturation to J1	0	33
2/1/72	I2♀	Gin	maturation to J1	0	33
2/1/72	J2♂	Even	maturation to sub	0	33
2/6–7/72	Ad♂	Dutch	emigration	−1	32

Table III (continued)

Date	Age-sex class	Individual	Event	Gain/ loss	New group size
2/8–9/72	Ad♂	Dutch	immigration	+ 1	33
2/21/72	J2♀	Slinky	immigration	+ 1	34
2/26/72	J2♀	Slinky	emigration	— 1	33
2/28/72	Ad♀	Ring	emigration	— 1	32
2/28/72	J1♀	Gin	emigration	— 1	31
2/28/72	Ad♂	Cowlick	emigration	— 1	30
3/3/72	Ad♀	Ring	immigration	+ 1	31
3/3/72	J1♀	Gin	immigration	+ 1	32
3/21/72	Ad♂	Peter	emigration	— 1	31
3/23/72	Ad♂	Peter	immigration	+ 1	32
4/22/72	I1♂	Inf. of Ring	birth (mother: Ad♀ Ring)	+ 1	33
4/24/72	I1♂	Inf. of Oval	still-birth (mother: Ad♀ Oval)	0	33
5/25/72	Ad♂	Cowlick	immigration	+ 1	34
6/13/72	Ad♂	Cowlick	emigration	— 1	33
6/19/72	I1♀	Inf. of Fluff	birth (mother: Ad♀ Fluff)	+ 1	34
6/25/72	Ad♂	Ivan	death or emigration	— 1	33
6/25/72	Ad♂	Cowlick	immigration	+ 1	34
7/1/72	Ad♂	Peter	emigration	— 1	33
7/4/72	Ad♂	Peter	immigration	+ 1	34
7/4/72	Ad♂	BJ	emigration	— 1	33
7/5/72	Ad♂	BJ	immigration	+ 1	34
7/8/72	Ad♂	Crest	immigration	+ 1	35
7/15/72	I1♀	Bell	maturation to I2	0	35
7/15/72	I1♀	Mindi	maturation to I2	0	35

Scan Samples

The study group was censused daily (table III); and the condition of each female's perineum was recorded every day. The data on females included size of sexual skin swelling on a scale of 1–20 (fig. 13), color of para-callosal skin [ALTMANN, 1970], presence or absence of vaginal bleeding, and presence or absence of residual ejaculate on the perineum. This technique of obtaining data may be considered scan sampling [ALTMANN, 1974] of group composition and of reproductive states of females. Also, the group was scanned repeatedly during the day and all fresh wounds noted.

Ad libitum *Samples*

Ad libitum sampling of behavior was done throughout the day and all observed occurrences of agonism, mounting, grooming, sexual presenting, and masturbation that were not recorded in the focal animal samples (see below) were entered in the *ad libitum* sample record. The behavior patterns of aggression and submission and the method of scoring agonistic bouts are described in chapter II, while description and scoring of other patterns is discussed below.

Ad libitum samples provided much of the information on dominance relationships in the study group, to be described in chapter II. It is not assumed that the *ad libitum* samples yielded a random sample of interaction partners or of rates of agonistic interaction. However, the assumption is made that the observed outcomes of agonistic interactions within each pair of animals constituted a random sample of all outcomes within that pair of animals during observation periods.

Table IV presents the hourly and monthly distribution of *ad libitum* sample time during this study. All other time spent working on some baboon-related project – habitat survey, censusing of other groups, etc. – is listed as 'general field time' at the bottom of table IV.

Focal Animal Samples

Perhaps the most important behavior samples for this study were obtained by the focal animal sampling technique [ALTMANN, 1974]. For each focal sample, the sample duration, starting time, and identity of the focal individual were specified in advance and were independent of the behavior of the animals. During each sample, every onset of a fixed set of behaviors (see below) done to, or by, the focal animal was recorded as was 'time-out', the amount of time that the focal animal was out of view. The assumption is made that the behavior of the focal animal while out of view was a random sample of that individual's behavior during sample periods. Therefore, the data record from focal animal samples is believed to contain a sample of behavior that is unbiased with respect to frequency of occurrence of each of the specified behaviors, partner choices in social behaviors, and outcome of agonistic bouts during sample periods. In actual practice, 97.4% of all scheduled minutes (N = 36,140 min) of focal animal samples were completed, and this fact clearly indicates the superb observation conditions that prevailed during this study. The focal animal typically was out of view for only a few seconds per sample, for example when passing behind a bush during a group progression.

Three focal animal samples of 30 min duration were obtained in each morning of observation and three 30-min samples were obtained in each

Table IV. Distribution of *ad libitum* sample time and general field time, in minutes

Hour beginning	1971					1972									Total
	Aug.	Sept.	Oct.	Nov.	Dec.	Jan.	Feb.	March	April	May	June	July	Aug.	Sept.	
0700	0	116	210	523	304	360	434	489	435	229	154	98	37	19	3,408
0800	586	1 116	1,349	1,411	1,236	1,561	1,075	1,500	1,708	1,758	1,428	1,296	932	140	17,096
0900	1,188	1,522	1,331	1,297	1,006	1,385	1,005	1,500	1,611	1,771	1,425	1,471	1,021	110	17,643
1000	1,066	1,379	992	1,078	809	1,290	870	1,419	1,468	1,683	1,379	1,223	667	60	15,382
1100	1,052	1,278	920	1,036	770	1,178	840	1,730	1,228	1,560	1,340	794	367	60	14,153
1200	247	846	928	971	738	1,085	840	1,185	1,092	1,560	1,195	804	233	60	11,784
1300	1,110	1,284	1,047	1,029	699	1,066	840	1,146	1,085	1,505	1,230	924	267	60	13,292
1400	1,043	1,451	952	1,044	720	1,115	840	1,140	1,140	1,410	1,260	1,042	255	60	13,472
1500	1,151	1,459	1,006	1,203	789	1,200	840	1,145	1,140	1,451	1,260	1,140	272	60	14,116
1600	1,215	1,651	1,189	1,391	1,117	1,272	832	1,260	1,144	1,470	1,284	1,173	407	60	15,465
1700	1,133	1,613	1,039	1,126	1,066	1,274	851	1,032	1,097	1,318	1,213	1,064	626	61	14,513
1800	67	308	84	42	35	72	38	27	0	11	0	34	89	20	827
Ad libitum sample time	9,858	14,023	11,047	12,151	9,289	12,858	9,305	13,573	13,148	15,725	13,168	11,063	5,173	770	151,151
General field time	1,669	1,970	2,748	2,487	2,459	1,947	2,258	1,682	1,414	1,630	1,225	2,696	3,214	283	27,682
Total	11,527	15,993	13,895	14,638	11,748	14,805	11,563	15,255	14,562	17,355	14,393	13,759	8,387	1,053	178,833
Total, h	192.1	266.6	231.6	244.0	195.8	246.8	192.7	254.3	242.7	289.3	239.9	229.3	139.8	17.55	2,980.6

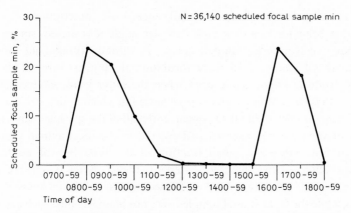

Fig. 3. Hourly distribution of focal animal sample time. Samples were concentrated during periods of greatest social activity within the group.

afternoon of observation. The focal sample on the first individual scheduled for each morning was begun on the 5-min interval after its descent from the sleeping trees, and descent usually occurred between 0800 and 0900 h. After-noon focal animal samples were begun at 1600 h each day. A 5-min rest was taken between the termination of one sample and the start of the next sample. These morning and afternoon sample periods corresponded to the times of the day when baboon social behavior usually occurred at the highest rate [ALTMANN and ALTMANN, 1970]. Figure 3 gives the hourly distribution of scheduled minutes of focal animal sample time.

Near midcycle (cf. chapter III) each female was sampled once in the morning and once again in the afternoon of the same day. All remaining focal animal samples, of the six that were taken each day, were allotted at random to adult or subadult males, though any individual male was sampled only once per day. Focal animal samples on cycling females were taken through-out the study period, while focal animal samples on adult and subadult males were begun on 7 March 1972 and continued until the end of the study.

Concurrent Sampling

Ad libitum sampling was not terminated when focal animal sampling was in progress. While taking focal animal samples, I inevitably also ob-served behaviors that were in the minimum set or otherwise of interest, but that did not directly involve the focal animal. For example, the focal animal might be seated and feeding passively while two other individuals presented, mounted, groomed, or aggressed just a few meters away. Though description of focal animal behavior always took precedence over description of non-focal animal behavior, use of a tape recorder made it relatively simple to record many such nonfocal animal behaviors without losing sight of the focal animal. Of course, data from focal animal samples were tabulated separately from *ad libitum* data, even when the latter were obtained concurrently with focal samples. For any time period in which both focal animal and *ad libitum* samples were taken concurrently, all of the specified behaviors of the focal animal were observed and entered in the focal animal sample record, whereas only some small fraction of the behaviors of nonfocal animals was observed and, therefore, entered in the *ad libitum* sample record. Doubtless, the *ad libitum* samples were biased against subtle, inconspicuous behaviors, while the focal animal samples were not biased against such events.

Specified Behaviors
All observed occurrences of a specified minimum set of behaviors, behavior set A, were recorded, either in the focal animal sample record, or the *ad libitum* sample record as appropriate, regardless of the age, sex, or identity of the participants in the interaction. This minimum set of behaviors included mounting, presenting, grooming, agonistic behaviors, and masturbation. Several additional behaviors, listed below as behavior set B, were recorded only in focal female samples.

Behavior set A. The minimum set of behaviors recorded during focal male, focal female, and *ad libitum* samples was as follows:
1. Mount: The onset of a mounting was recorded whenever one animal of either sex, in quick succession, grasped the pelvis of another animal of either sex, elevated his forequarters over the hind end of the other animal, and clasped the other animal's ankles with his feet (fig. 4, 6). The mounter thus supported his entire weight on the pelvis and hindlimbs of the mounted animal. Incipient or partial mountings were a specified behavior only in focal female samples (see behavior set B below).
Mountings commonly occurred without intromission, pelvic thrusting, or ejaculation and thus not all mountings were copulations. A male-female mounting was considered a copulation only if either the male gave an ejaculatory pause, or the female had fresh ejaculate on her perineum immediately after the mounting, or both. Thus a copulation, as defined here, is a mounting culminating in ejaculation.
2. Present hindquarters: The onset of a presentation was recorded whenever one individual approached to within three meters of another and

Fig. 4. Mount, present, grasp pelvis, and perineal inspection. *A* Adult male Stubby mounted female Fluff. Note that male is completely supported by female's ankles and pelvis. *B* Adult female Alto presented to adult male BJ. BJ grasped Alto's pelvis and visually inspected her perineum; Alto reached back with her left hand and touched BJ's genitals (significance unknown).

turned his or her hindquarters directly toward the face of the other individual (fig. 4). No particular response by the recipient was necessary for a present to be scored, though the response to a presentation was always noted.

3. Groom: The onset of a grooming was recorded whenever one individual parted his own fur or that of another animal and picked or scraped at the hair or skin. The onset of bouts of one individual continuously grooming himself or another were scored in this category, rather than the occurrence of individual strokes of the hand of the groomer. For example, if animal A groomed animal B for ten hand strokes, and then B groomed A for 100 hand strokes, only two onsets of grooming were recorded.

4. Agonistic behaviors: All occurrences of overt aggression (fighting) between two or more individuals, herding of one individual by another, and more subtle spatial displacements of one animal by another were recorded under this category. A more complete description of the behaviors of aggression and submission and a method for scoring bouts of agonistic behavior is given in chapter II.

5. Masturbation: Masturbation was recorded whenever a male manipulated his penis. Ejaculation was not necessary for masturbation to be recorded, though masturbation to ejaculation was common. No behavior that could be interpreted as female masturbation was observed in this study.

Behavior set B. In focal female samples, all occurrences of behaviors in set A (above) and of the following behaviors were recorded:

6. Inspect perineum: An onset of perineal inspection was recorded whenever one individual touched, sniffed, or stared directly at the perineum of another individual (fig. 4).

7. Follow: One occurrence of following was recorded each time one animal walked directly after another in a movement through the group (fig. 5). The bout of following was considered terminated when the follower ceased walking for approximately 15 sec or longer.

8. Grasp pelvis: One occurrence of pelvis grasping was scored whenever one individual used one or both hands to grip or touch the pelvis of another individual, but did not proceed any further into the mounting position (see above) (fig. 4). This behavior often occurred in response to a presentation of the hindquarters (see above), but was always recorded whether it occurred in a presentation or nonpresentation context.

9. Incipient mount: One occurrence of incipient mounting was scored whenever one animal of either sex grasped the pelvic region of another individual of either sex and rapidly elevated his forequarters over the hind

Fig. 5. Follow and incipient mount. *A* Adult male Peter followed female Oval, estrus 10 swelling. *B* Peter incipient mounted Oval, estrus 10 swelling.

end of the other animal, but did not proceed any further into the mounting position (see above) (fig. 5).

10. Consortship: At the termination of each focal female sample, the identity of the male consorts, if any, of the sampled female was recorded. A consortship was scored when a male and female showed continual attention to, and awareness of, each other's location, and when one or both animals moved so as to maintain close spatial proximity with the other. Although in this study consortship was a subjective evaluation of the part of the observer, it seems clear that the consort relationship could be objectively defined in terms of persistent nearest neighbor relationships.

The behaviors in set B (excluding consortship evaluation) and the behaviors mount, present, and groom from set A taken together form an exclusive, but not exhaustive, set of baboon social behaviors. Thus, no instances of one animal simultaneously grooming and mounting another, etc., were observed. However, the agonistic behaviors, or acts, described in chapter II form an exhaustive, but not exclusive set of baboon agonistic behaviors. Thus, one animal could simultaneously raise his eyebrows and ground-slap (two aggressive acts) at a second individual, or simultaneously present to (a nonagonistic behavior), and grimace at (a submissive behavior) the second animal.

Furthermore, in analyzing dominance, I distinguish between bouts, or continuous sequences of behaviors, and the individual behaviors themselves, the component acts of a bout. A bout also may be considered a set of contiguous behaviors, or acts, and a behavior may be considered one element of a bout. In the present scoring system only one bout of behavior could be in progress within a particular pair of animals at any one time, though one individual could be participating simultaneously in bouts of behavior with more than one individual. Thus, individual A could not be simultaneously participating in, say, a grooming bout and an agonistic bout with individual B. If A groomed, then bit, B, the grooming bout was terminated by the act of biting, i.e. by the onset of an agonistic bout. However, individual A could be participating in a grooming bout with individual B while simultaneously a participant in an agonistic bout with a third individual, C. In fact, the baboons only very rarely violated any of these scoring conventions. In the present study, every bout or sequence of behaviors in which at least one agonistic behavior occurred was classed as a bout of agonistic behavior and analyzed for information on dominance regardless of the number of nonagonistic behaviors that also may have occurred in the bout. The analysis of these bouts of agonistic behavior will be discussed in detail in the following chapter.

A

B

Fig. 6. Mount by immature male and mount by subadult male. *A* Infant-2 male Major mounted estrous female Fluff. *B* Subadult male Ben mounted estrous female Ring while Ring's adult male consort was temporarily absent. Note that Ben glanced furtively about during the mount.

II. Dominance Relationships in the Study Group

Overview

The study of agonistic behavior in animals is one aspect of the description of social order within their groups. Dominance usually is considered an aspect of agonistic behavior and is inferred whenever one animal 'is able to chastise another with impunity' [KLOPFER and HAILMAN, 1967, p. 142]. The development of the concept of dominance and its original close relationship to fundamental biological problems is outlined by COLLIAS [1944]. The articles in CARTHY and EBLING [1964] review the implications of dominance and aggression in other animals for our understanding of human behavior.

The study of aggression in primates has become almost synonymous with the study of dominance hierarchies, their linearity and stability through time [MASLOW, 1936; KAWAI, 1965; KAWAMURA, 1965; SADE, 1967; MISSAKIAN, 1972]. As RIOCH [1967, p. 116] has pointed out, the notion of hierarchy has become formalized and often is 'thought of as literally referring to some occult force that directs behavior'. A linear hierarchy of agonistic dominance, in particular, is the *result*, rather than the *cause*, of consistency and transitivity of agonistic relations within pairs of animals: The dominant individual in a pair of animals is the one that consistently directed attack or threat behaviors toward the subordinate, who in turn consistently responded with behaviors of submission.

Studies that focus on hierarchy, rather than consistency of paired relationships, run the strong risk of neglecting or ignoring important information about the behavior of the study animals. Reversals, i.e. agonistic bouts in which the presumed subordinate defeated the presumed dominant animal, are often viewed as minor perturbations of some stable underlying social structure represented by the dominance hierarchy. However, when consistency of relationship is the focus of a behavioral study, reversals become data for consideration and evaluation. Such 'minor perturbations' in the hierarchy may actually be the result of fundamental biological processes, such as sexual cycling or maturation.

Additionally, criteria of dominance and the kinds of social interactions that are considered to yield information about dominance vary greatly among different observers of primates [GARTLAN, 1968]. In the literature, dominance-yielding interactions have included all bouts of agonistic behavior [SADE, 1967], a subset of agonistic bouts, such as overt fights, but not more subtle aggression, or certain nonagonistic behaviors such as leading the group or protecting infants [DEVORE, 1965]. To make matters even more confusing, most observers have failed to state explicitly what behaviors were used as criteria of dominance.

In the following discussion of dominance relationships in the study group, the consistency of outcome of successive bouts of agonistic behavior within pairs of animals is evaluated. The kinds of bouts which are considered agonistic and the method of scoring such bouts will be stated explicitly.

Methods

Behaviors of Aggression and Submission

Table V lists the behaviors of aggression and submission in yellow baboons *(Papio cynocephalus)* and comparable behaviors that have been reported in other species of the genus *Papio* (fig. 7). Every observed occurrence of one or more of these behaviors was recorded either in the *ad libitum* record or in the record of the ongoing focal animal sample, as appropriate. These behaviors were first recognized in sequences of behavior which culminated in overt, contact aggression, such as one animal biting or soundly hitting another. A behavior was assigned to the list of submissive behaviors (table V B) if it was given by animals toward whom overt aggression was directed. Similarly, a behavior was assigned to the list of aggressive behaviors (table V A) if it was given by the individual who performed the act of aggression. In baboons, the recipient of an act of overt aggression often gives a brief counter-case, or counter-attack, to the aggressor. However, the other behaviors given by the counter-chaser during the counter-chase itself, particularly facial expression and tail position, usually are submissive, and hence I have also assigned counter-chase to the list of submissive behaviors.

Table V does not list the vocal behaviors which accompany aggression since no analysis of vocalizations was made during this study. In general, the vocalizations of the submissive animal were high-pitched screeches, squeals, and cackles, while the vocalizations of the aggressor were relatively low-pitched grunting sounds.

Table V. Behaviors of aggression and submission in baboons (genus *Papio*)

Behavior	Description	Behaviors in other baboon studies	
		HALL and DE VORE [1965]	KUMMER [1957, p. 28, table IV; 1968; p. 180]

A. Behaviors of Aggression

Behavior	Description	HALL and DE VORE [1965]	KUMMER [1957, p. 28, table IV; 1968; p. 180]
1. Stared	obvious	staring (directed) (tables 3–6, p. 92)	staring [1968]
2. Raised brow	eyebrows raised, frontalis muscle tensed (fig. 7)	eyebrow-raising (tables 3–6, p. 92)	?contracting brows [1957] Raising brows [1968]
3. Pulled ears back	auricle of ear pressed against side of head (fig. 7)	ear flattening (tables 3–6, p. 92; tables 3–6, p. 95)	—
4. Gave open-mouth face	lower jaw is dropped open, mouth assumes oval shape, corners of mouth not retracted (fig. 7)	—	opening mouth [1968]
5. Bobbed head and thorax	abrupt, rapid raising and lowering of head and trunk; body may show a forward movement component	jerking of head down and forward (tables 3–6, p. 92)	protruding head [1968]
6. Ground-slapped	palm of hand or hands struck against ground, often with audible noise	slapping ground with hand (tables 3–6, p. 92)	beating the ground [1957] slapping ground [1968]
7. Lunged	leaping or jumping toward another individual; not sharply distinguished from above behavior or from chasing	charging run (tables 3–6, p. 93)	feigned attack [1957] lunging at partner [1968]
8. Gave exaggerated chewing motion of the jaws	repeated chewing or grinding movements of jaws, but with extreme lateral excursion and copious salivation; tongue and teeth produce noises that are sometimes audible at greater than 50 ft	tooth-grinding (tables 3–6, p. 92)	pumping cheeks with chewing motions [1968]
9. Yawned	a directed behavior, as opposed to casual yawning when tired, yawn given (1) with teeth covered by lips entirely,	yawning (tables 3–6, p. 93)	?opening mouth, high intensity [1968]

Table V (continued)

Behavior	Description	Behaviors in other baboon studies	
		HALL and DeVore [1965]	KUMMER [1957, p. 28, table IV; 1968, p. 180]
	(2) with canines only exposed at apex of yawn, or (3) with teeth and gums fully exposed at apex of yawn		
10. Rubbed muzzle against substrate	muzzle and chest rubbed on ground or tree limbs; often accompanied by the two patterns immediately above	rubbing against ground (tables 3–7, p. 97)	—
11. Chased	not obviously distinguishable from counter-chase except on basis of accompanying behavior (table V B); see text for description (fig. 7)	charging run (tables 3–6, p. 93)	chases [1968]
12. Hit, grabbed or push	one animal attempted to make, or actually made, physical contact with another	grappling with hands (tables 3–7, p. 97); slapping (tables 3–7, p. 97)	pulling closer [1968] shoving away [1968]
13. Bit or nipped	obvious (fig. 7)	biting (tables 3–7, p. 97); biting gently at nape of neck (tables 3–7, p. 97)	biting in the nape of neck [1957] biting on shoulder [1968]

B. Behaviors of submission

1. Glanced repeatedly	multiple sideways glances at an approaching animal	sideways jerking glances (tables 3–6, p. 93)	—
2. Avoided eye contact	head turned aside, gaze fixed on ground	looking away (tables 3–6, p. 93)	—
3. Cowered	lateral flexion of the spine, often from a seated position	shoulder shrugging (tables 3–6, p. 94)	?bending elbows and knees while standing [1968] bending legs [1967]
4. Walked away or ran away	obvious. Often preceded by a quick startle movement (fig. 7)	—	Walking away, avoiding, escaping [1968] flight [1957]

Table V (continued)

Behavior	Description	Behaviors in other baboon studies	
		HALL and DE VORE [1965]	KUMMER [1957, p. 28, table IV; 1968; p. 180]
5. Crouched toward ground, or gave fear paralysis	limbs are flexed and adducted, animal leans forward or lies on ground motionless (fig. 7)	body prone to ground, animal rigid (tables 3–6, p. 94)	pressing down [1957] crouching [1968]
6. Grimaced, or gave cackle-face	corners of mouth retracted and teeth clenched during grimace; cackle-face is basically same expression, but teeth are unclenched allowing production of 'cackle' vocalization (fig. 7)	grin (tables 3–6, pp. 93 and 95)	baring teeth with closed jaws [1968]
7. Gave tail-up	tail held upright in the air (fig. 7)	tail erect (tables 3–6, p. 94)	tail high [1957]
8. Gave counter-chase	not obviously distinguishable from chase (table V A); see text for description	charging run (tables 3–6, p. 93)	—

Bouts of Agonistic Behavior. Definition and Scoring

Bouts of agonistic behavior were scored as events and the onset of each bout was taken as the event to be scored. The onset of a bout of agonistic behavior was scored at the first occurrence of any one of the behaviors in table V, and the bout was terminated only when all participants engaged in some behavior other than continued aggression or submission toward the other participants. As noted in chapter I, only one bout of agonistic behavior could be in progress within a particular pair of animals at any time, though an individual could be participating simultaneously in bouts of agonistic behavior with more than one other individual.

Of course, not all behaviors listed in table V were given in every agonistic bout, nor did all bouts of such behavior culminate in overt aggression, such as hitting or biting. Thus, the definition of agonistic bout used in this study covered a wide variety of sequences of social behavior, ranging from subtle spatial displacements or supplantation, in which one animal merely moved quickly aside from another, to overt fighting in which physical contact and wounding occurred. Similarly, nonagonistic behaviors often occurred within bouts of otherwise agonistic behavior, for example when one individual simultaneously grimaced and presented to another. However, the consistency of outcome of successive agonistic bouts within a pair of individuals showed little dependence on the apparent intensity of the agonistic behaviors in their bouts.

Outcome Determination

Every observed agonistic bout was analyzed as a potential source of dominance information but, as will be explained below, not all bouts actually provided such information. Bouts of only nonagonistic behaviors were not considered as sources of dominance information, but were analyzed later as possible correlates of agonistic dominance. Agonistic bouts were divided into 'decided bouts', i.e. those for which a 'winner' and 'loser' could be determined by the criteria given below, and into 'undecided bouts', i.e. those for which a winner and a loser could not be determined. In fact, over 99% of all agonistic bouts recorded in focal female samples (N = 656 bouts), and over 97% of the agonistic bouts recorded in the focal male samples (N = 1,086 bouts) were decided by the criteria described below.

Winner-loser determinations were made on the basis of the behaviors given during agonistic bouts. A winner and a loser were determined in an agonistic bout only when one animal directed one or more submissive behaviors, and no aggressive behaviors, toward a second animal in response to aggressive behaviors (or any other set of nonsubmissive behaviors) from the second animal. The individual who gave the submissive behaviors was considered the loser of the bout, and the individual who gave only aggressive and/or other nonsubmissive behaviors was considered the winner of the bout.

By the above criteria, one animal, A, would be scored as the winner in a bout of agonistic behavior with a second animal, B, if A casually walked toward B and B grimaced at A, or if A presented to B and B grimaced at A, or if A bit B and B grimaced in response, or if B approached and grimaced into A's face as A sat feeding calmly. In each such decided dyadic agonistic bout, only one winner and one loser was recorded. The scoring of multiple individual agonistic bouts will be discussed in the next section of this chapter.

When two individuals chased or grappled with each other and then parted with neither individual giving submissive behaviors, the agonistic bout was scored as undecided. Such undecided agonistic bouts were most common between immature males and adult females, or between two adult males undergoing a period of inconsistent dominance relationship (see below). Also occasionally a higher ranking adult female gave 'tail up', a submissive behavior (table V B), to another lower ranking female who was carrying a black infant (I 1); however, the lower ranking female usually responded by sitting rigidly and avoiding eye contact with the other female. Thus, these bouts of females giving submissive behaviors mutually were also scored as undecided. Finally, if one individual threatened another and the second individual literally ignored the threat of the first, an undecided agonistic bout was scored, but this was an extremely rare event.

Multi-Individual Agonistic Bouts

Triadic, or tripartite, agonistic bouts resulted when one individual simultaneously aggressed against, or was aggressed against by, two other individuals, or when one individual aggressed against a second individual who simultaneously aggressed against yet a third individual. Triadic agonistic bouts and bouts with more than three participants were divided into their component dyads and scored according to the procedure outlined above for dyadic agonistic bouts. For example, if one animal, A, simultaneously attacked two other animals, B and C, or vice versa, the triadic bout was divided into A-C, A-B, and B-C component dyads and then a winner and loser scored, if possible, within each of these pairs. Again for example, in the case of A simultaneously attacking B and C, both of whom responded with submissive behaviors, A was scored as the winner in a dyadic agonistic bout with B and as the winner in a dyadic agonistic bout with C. However, no outcome was recorded in the B-C dyad unless B and C specifically exchanged agonistic behaviors.

In fact, multiple individual agonistic bouts were rare. Less than 2% of the agonistic bouts recorded in the focal animal samples (N = 1,742 bouts) actually involved more than two individuals. When these multi-individual agonistic bouts were broken into their component dyads and scored as described above, no modification of paired dominance relationships as a

---

Fig. 7. Behaviors of aggression and submission in yellow baboons, *Papio cynocephalus. A* Adult male Stubby (far right), eyebrows raised, ears pulled back, and with open mouth face, chased adult male Dutch (far left) who ran away with tail raised in the air. Note that Dutch carried a black infant ventrally during the agonistic bout. *B* Juvenile-2 male Russ crouched to ground in fear paralysis, grimaced, screamed, and raised tail in response to attack from another individual not shown in photograph. *C* Adult female Mom, with eyebrows raised, grabbed and bit adult female Judy who exhibited mild fear paralysis and raised tail slightly. The bite did not actually result in a wound.

A

B

C

result of participation in such 'coalitions' was found. It was my impression, however, that two low ranking adult males in coalition were more likely to counter-chase a higher ranking adult male than was either low ranking male alone.

Multiple individual agonistic bouts have been reported to occur quite frequently in some groups of anubis *(Papio anubis)* and chacma *(P. ursinus)* baboons [HALL and DEVORE, 1965]. Specifically, multi-individual agonistic interactions among adult males were particularly frequent in DEVORE's Songora Ridge (S.R.) Troop, Nairobi Park, Kenya, when only a single estrous female was present [DEVORE, 1965]. This troop also had a relatively high male:female sex ratio in adults compared to other troops in the Nairobi Park population. Tripartite or three individual agonistic bouts also occurred frequently in the captive group of hamadryas baboons *(P. hamadryas)* observed by KUMMER [1957], but such multi-individual agonistic bouts were only rarely observed in wild populations of hamadryas [KUMMER, 1968]. Thus, while multi-individual agonistic bouts may be an important feature of the social organization of a few specific baboon groups, it seems likely that in general such interactions will be of little importance in understanding dominance relations in baboons.

Consistency of Outcome and Dominance

If one animal, A, was the winner in two successive decided agonistic bouts with another individual, B, then the dominance relationship between A and B was defined as 'consistent', with A dominant to B, on all days intervening between the days of the two decided bouts and also on the day of the second of the two bouts. Similarly, if A and B were each the winner in one of two successive decided agonistic bouts, then the A-B dominance relationship was defined as 'inconsistent', with neither A nor B dominant or subordinate, on all days intervening between the days of the two decided bouts and also on the day of the second of the two bouts. In fact, it was, during these periods of inconsistent dominance relationship that most undecided agonistic bouts occurred.

If two or more inconsistent bouts occurred on the same day in a pair of individuals, their dominance relationship was considered to be inconsistent for the entire day. Thus, by this scoring system, the minimum unit of time that the dominance relationship within a pair of individuals could be inconsistent (or consistent) was one full day. Also, for those periods that I was absent from the field for an entire day or longer, dominance relationships within all pairs of individuals in the study group have been assumed to have

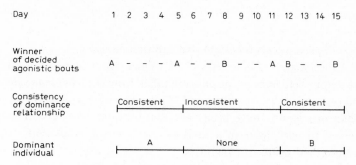

Fig. 8. Hypothetical sequence of agonistic bout outcomes. See text for further explanation.

remained as they were prior to my absence; however, these assumed relationships have been clearly marked as such in all tables and figures that follow.

As an illustration of this procedure, figure 8 presents a hypothetical sequence of outcomes of successive decided agonistic bouts between animals A and B for a 15-day period. An 'A' in line 2 indicates a bout in which A was the winner, a 'B' indicates a bout in which B was the winner, and a dash indicates that no decided agonistic bout was observed on that day. In the sequence outlined in figure 8, the dominance relationship between A and B is defined as consistent, with A dominant to B, on days 2 through 5, and defined as consistent, with B dominant to A, on days 13 through 15. On days 6 through 12, the relationship between A and B is defined as inconsistent, with neither A nor B dominant.

Under the strict definition of consistency of relationship used in this analysis, one decided agonistic bout within a pair of individuals that disagreed with the outcome of the immediately preceding decided bout was sufficient to define a period of inconsistency of relationship. Such an inconsistent outcome is equivalent to a reversal on the familiar dominance matrix.

Dominance Rank

Finally, the dominance rank of an individual at any time was equal to (n-r), where n equaled the number of individuals in the group or subgroup under evaluation, and r equaled the number of other individuals to whom he was consistently dominant. For example, if one individual was consistently dominant to six other individuals in a group of ten animals, he was assigned the rank number of four. The use of this method of rank number assignment allowed two or more individuals to be assigned the same rank number and thus accommodated triangular or other nonlinear dominance relationships.

Results

Female Dominance Relationships. Ad Libitum *Samples*

Table VI is a matrix of outcomes of all decided agonistic bouts in which both participants were female; the data are taken from the *ad libitum* sample records. Three females, Twisty, New, and Slinky, who were with Alto's group for less than one month, have been deleted from table VI. These females were observed to participate in a total of ten agonistic bouts with other females before their death or emigration.

Table VI. Outcome of agonistic bouts between females, *ad libitum* samples. This table lists the observed outcome of all decided agonistic bouts between females as recorded in the *ad libitum* sample record; winner of the bout is indicated by the row of the cell and the loser by the column of the cell

Winner	Loser																
	Adult ♀♀											J2♀, Fem	J1♀♀		12♀♀		11♀, Fluff's inf.
	Skinny	TT	Alto	Mom	Lulu	Fluff	Preg	Scar	Oval	Judy	Ring		Vee	Gin	Bell	Mindi	
Adult ♀♀																	
Skinny	–	3	2	2	2	3	2	1	4	2	3	1	1	0	0	0	0
TT	0	–	37	26	21	27	26	20	19	24	15	16	7	11	1	0	0
Alto	0	0	–	22	39	34	31	23	45	21	70	17	4	2	0	0	0
Mom	0	0	0	–	39	38	61	30	34	29	20	8	1	1	0	0	0
Lulu	0	0	0	0	–	26	43	35	35	15	12	9	4	6	0	0	0
Fluff	0	0	0	0	0	–	27	27	45	19	24	8	3	2	0	0	0
Preg	0	0	0	0	0	0	–	19	34	17	8	17	5	4	1	0	0
Scar	0	0	0	0	0	0	1	–	58	34	16	16	1	10	0	0	0
Oval	0	0	0	0	0	0	0	0	–	43	19	37	6	6	0	0	0
Judy	0	0	0	0	0	0	0	0	0	–	12	9	4	7	0	0	0
Ring	0	0	0	0	0	0	0	0	0	0	–	7	0	6	0	0	0
J2♀																	
Fem	0	0	0	0	0	0	0	0	0	4	1	–	5	36	0	0	0
J1♀																	
Vee	0	0	0	0	0	0	0	0	0	0	0	0	–	10	0	0	0
Gin	0	0	0	0	0	0	0	0	0	0	0	0	0	–	0	0	0
12♀♀																	
Bell	0	0	0	0	0	0	0	0	0	0	0	0	0	0	–	0	0
Mindi	0	0	0	0	0	0	0	0	0	0	0	0	0	0	0	–	0
11♀																	
Fluff's inf.	0	0	0	0	0	0	0	0	0	0	0	0	0	0	0	0	–

Table VI shows strong consistency of paired relationship between the females in Alto's group: only six (0.4%) reversal outcomes were recorded in a total of 1,638 decided agonistic bouts among females. Five of these reversals involved J2 female Fem interacting with adult females Ring or Judy. Maturation of J2 females to adulthood probably produces much of the inconsistency in female dominance relationships [cf. SADE, 1972].

The one remaining reversal occurred between adult females Preg and Scar and will be analyzed in detail in chapter V. However, this one reversal among adult females constituted less than 0.1% of all outcomes of decided agonistic bouts among adult females in *ad libitum* samples (N = 1,344 bouts) and constituted 5% of all outcomes of decided agonistic bouts within the Preg-Scar pair (N = 20 bouts).

Focal Female Samples

Table VII presents the outcome of decided agonistic bouts between females as recorded in the focal female samples. A total of 364.88 h of focal female sampling were completed during the study period (table XXI). Focal female samples were taken only on cycling (and therefore adult), nonpregnant females from the seventh day prior to the expected day of deturgescence of the sex skin through the fourth day after the onset of deturgescence. In the analyses that follow, this interval will be referred to as 'midcycle'; and in fact ovulation occurs within this portion of the baboon menstrual cycle (chapter III). As a result of the above sampling restrictions, at least one of the two females in all of the agonistic bouts that were recorded in focal female samples was in midcycle. Consequently, if neither member of a pair of females cycled during the study period, then no information on dominance relationships within that pair could be provided by the focal female samples that were taken, and the outcome of agonistic bouts within pairs of females neither of whom cycled during the study period are considered to be undefined (*a priori* zero) and have been indicated by an asterisk in the appropriate cell in table VII.

Altogether 656 agonistic bouts in which at least one participant was a female in midcycle were recorded during focal female samples. Five (0.7%) of these interactions were undecided by the criteria listed above. All undecided agonistic bouts in the focal female samples were between an adult female and a J2 male, again demonstrating that maturation of juveniles was a major factor producing inconsistency of dominance relationship in the study group. Because the systematic focal animal samples provided an unbiased sample of outcomes of agonistic bouts, at least during midcycle, the results in table VII cannot be accounted for by bias toward specific outcomes. Beyond that, they are in close agreement with respect to the rank ordering of adult female and the strong consistency of female-female dominance relationships indicated by the *ad libitum* data (table VI).

Male-Female Dominance Relationships

Tables VIII and IX summarize the observed outcomes of decided agonistic bouts between adult females and males of all ages as recorded in the *ad libitum*, focal male, and focal female samples. Table VIII shows only

Table VII. Outcome of agonistic bouts between females, focal female samples, listing the observed outcome of all decided agonistic bouts between females as recorded in the focal female sample record. An asterisk indicates a cell that was *a priori* zero due to sampling procedures. See text for further explanation. Other conventions as in table VI

Winner	Loser											J2♀	J1♀♀		I2♀♀		I1♀
	Adult ♀♀											Fem					Fluff's inf.
	Skinny	TT	Alto	Mom	Lulu	Fluff	Preg	Scar	Oval	Judy	Ring		Vee	Gin	Bell	Mindi	
Adult ♀♀																	
Skinny	–	*	0	*	1	2	*	*	1	1	0	*	*	*	*	*	*
TT	*	–	7	*	15	5	*	*	6	6	1	*	*	*	*	*	*
Alto	0	0	–	5	30	12	14	3	14	18	3	0	0	0	0	0	0
Mom	*	*	0	–	7	4	*	*	10	31	2	*	*	*	*	*	*
Lulu	0	0	0	0	–	14	27	14	22	24	12	17	2	3	0	0	0
Fluff	0	0	0	0	0	–	5	1	6	4	6	2	0	1	0	0	0
Preg	*	*	0	*	0	0	–	*	16	16	1	*	*	*	*	*	*
Scar	*	*	0	*	0	0	*	–	12	16	1	*	*	*	*	*	*
Oval	0	0	0	0	0	0	0	0	–	12	2	4	1	1	0	0	0
Judy	0	0	0	0	0	0	0	0	0	–	9	1	1	3	0	0	0
Ring	0	0	0	0	0	0	0	0	0	0	–	0	0	0	0	0	0
J2♀																	
Fem	*	*	0	*	0	0	*	*	0	0	0	–	*	*	*	*	*
J1♀♀																	
Vee	*	*	0	*	0	0	*	*	0	0	0	*	–	*	*	*	*
Gin	*	*	0	*	0	0	*	*	0	0	0	*	*	–	*	*	*
I2♀♀																	
Bell	*	*	0	*	0	0	*	*	0	0	0	*	*	*	–	*	*
Mindi	*	*	0	*	0	0	*	*	0	0	0	*	*	*	*	–	*
I1♀																	
Fluff's inf.	*	*	0	*	0	0	*	*	0	0	0	*	*	*	*	*	–

agonistic bouts in which an adult male defeated an adult female but, in fact, all 1,801 decided agonistic bouts between adult males and adult females were won by the adult males. In other words, all adult males were consistently dominant to all adult females during the study period. It will be recalled,

Table VIII. Outcome of agonistic bouts between adult males and adult females, all sample records. The first entry in each cell shows the observed outcome of all agonistic bouts between adult males and adult females as recorded in *ad libitum*, focal male, and focal female sample records combined. The second entry in each cell shows all outcomes recorded in focal male samples and the third entry all outcomes recorded in the focal female samples. Other conventions as in tables VI and VII

Winner, adult males	Loser, adult females										
	Skinny	TT	Alto	Mom	Lulu	Fluff	Preg	Scar	Oval	Judy	Ring
BJ	0 (combined)	25	25	23	42	20	18	13	25	7	20
	* (focal ♂)	2	2	1	0	2	3	1	1	0	3
	* (focal ♀)	*	1	*	25	1	*	*	6	1	0
Peter	0	14	31	15	20	35	7	6	11	2	17
	*	1	6	2	3	11	4	1	0	0	2
	*	*	2	*	4	1	*	*	4	1	1
Ivan	2	73	165	33	60	97	57	28	19	43	58
	*	8	9	3	2	4	4	5	0	0	4
	*	*	67	*	33	41	*	*	2	27	9
Crest	*	13	8	6	24	17	5	4	12	33	9
	*	0	0	0	0	0	0	0	0	0	0
	*	*	0	*	22	2	*	*	3	24	0
Stubby	4	30	32	7	12	24	19	11	34	19	27
	*	3	2	3	0	1	2	2	2	2	3
	*	*	13	*	6	8	*	*	8	7	0
Dutch	0	18	13	10	22	7	6	3	47	72	11
	*	1	0	0	1	0	1	1	7	0	2
	*	*	3	*	13	0	*	*	28	60	1
Sinister	0	8	6	5	1	4	7	3	12	5	8
	*	2	3	1	0	3	2	1	2	3	1
	*	*	0	*	0	1	*	*	0	0	0
Max	1	18	19	17	10	13	14	4	12	8	4
	*	7	2	2	0	1	7	1	4	1	0
	*	*	2	*	2	0	*	*	2	2	1
Cowlick	1	6	6	7	3	11	2	4	5	3	4
	*	0	0	1	0	0	1	0	4	0	1
	*	*	2	*	2	1	*	*	1	2	0

Table IX. Outcome of all agonistic bouts between adult females and nonadult males, all sample records. The first entry in each cell gives the total number of outcomes in *ad libitum*, focal female, and focal male sample records combined. The second entry in each cell is the number of outcomes recorded in the focal female samples and the third entry the number of outcomes recorded in focal male samples. All other conventions as in tables VI and VII

A. Outcome of all decided agonistic bouts in which a nonadult male defeated an adult female

Winner	Loser, adult ♀♀										
	Skinny	TT	Alto	Mom	Lulu	Fluff	Preg	Scar	Oval	Judy	Ring
Subadult ♂♂											
Ben	2 (combined)	24	25	17	28	12	15	3	21	15	6
	* (focal ♀)	*	1	*	6	4	*	*	3	7	3
	* (focal ♂)	5	3	3	0	2	2	1	6	3	3
Even	3	48	57	27	8	31	36	19	34	18	14
	*	*	1	*	0	8	*	*	5	4	1
	*	11	7	3	0	4	2	6	5	2	3
J2 ♂♂											
Stiff	0	3	11	2	0	2	9	3	1	2	6
	*	*	0	*	0	0	*	*	0	1	0
	*	*	*	*	*	*	*	*	*	*	*
Stu	0	0	2	3	0	4	1	3	12	2	5
	*	*	0	*	0	0	*	*	0	0	0
	*	*	*	*	*	*	*	*	*	*	*
Red	0	0	6	9	2	16	16	11	12	15	11
	*	*	1	*	0	1	*	*	0	2	3
	*	*	*	*	*	*	*	*	*	*	*
Russ	0	0	2	2	0	8	6	7	4	2	6
	*	*	0	*	0	0	*	*	0	1	0
	*	*	*	*	*	*	*	*	*	*	*
J1 ♂♂											
Swat	0	0	0	0	0	0	2	2	0	2	4
	*	*	0	*	0	0	*	*	0	0	1
	*	*	*	*	*	*	*	*	*	*	*
Major	0	0	0	0	0	0	0	0	0	0	0
	*	*	0	*	0	0	*	*	0	0	0
	*	*	*	*	*	*	*	*	*	*	*
Spot	0	0	0	0	0	0	0	0	1	0	0
	*	*	0	*	0	0	*	*	1	0	0
	*	*	*	*	*	*	*	*	*	*	*
12 ♂♂											
Kub	0	0	0	0	0	0	0	0	0	0	0
	*	*	0	*	0	0	*	*	0	0	0
	*	*	*	*	*	*	*	*	*	*	*
Dogo	0	0	0	0	0	0	0	0	0	0	0
	*	*	0	*	0	0	*	*	0	0	0
	*	*	*	*	*	*	*	*	*	*	*
1 1♂											
I-Ring	*	0	0	0	0	0	0	0	0	0	0
	*	*	0	*	0	*	*	*	0	0	*
	*	*	*	*	*	*	*	*	*	*	*

Table IX (continued)

B. Outcome of all decided agonistic bouts in which an adult female defeated a nonadult male

Winner	Loser											
	Subadult ♂♂		J2♂♂				J1♂♂			I2♂♂		I1♂, I-Ring
	Ben	Even	Stiff	Stu	Red	Russ	Swat	Maj.	Spot	Kub	Dogo	
Adult ♀♀												
Skinny	0 (combined)	0	0	0	0	2	2	1	0	0	0	*
	* (focal ♀)	*	*	*	*	*	*	*	*	*	*	*
	* (focal ♂)	*	*	*	*	*	*	*	*	*	*	*
TT	0	0	0	6	4	4	3	0	13	2	0	0
	*	*	*	*	*	*	*	*	*	*	*	*
	0	0	*	*	*	*	*	*	*	*	*	*
Alto	0	0	6	5	1	12	12	0	16	4	2	0
	0	0	2	1	1	1	2	0	4	0	0	0
	0	0	*	*	*	*	*	*	*	*	*	*
Mom	0	0	0	0	0	2	4	0	2	2	0	0
	*	*	*	*	*	*	*	*	*	*	*	*
	0	0	*	*	*	*	*	*	*	*	*	*
Lulu	0	15	19	7	8	6	7	1	0	2	0	0
	0	5	5	2	2	1	0	0	0	1	0	0
	0	0	*	*	*	*	*	*	*	*	*	*
Fluff	0	0	0	0	0	0	1	0	1	0	0	0
	0	0	0	0	0	0	0	0	0	0	0	*
	0	0	*	*	*	*	*	*	*	*	*	*
Preg	0	0	0	0	0	0	1	0	6	4	1	0
	*	*	*	*	*	*	*	*	*	*	*	*
	0	0	*	*	*	*	*	*	*	*	*	*
Scar	0	0	1	0	0	0	0	0	0	3	0	0
	*	*	*	*	*	*	*	*	*	*	*	*
	0	0	*	*	*	*	*	*	*	*	*	*
Oval	0	0	0	1	1	1	3	2	0	2	0	0
	0	0	0	0	0	0	0	0	0	0	0	0
	0	0	*	*	*	*	*	*	*	*	*	*
Judy	0	0	0	1	1	0	0	0	2	0	0	0
	0	0	0	0	0	0	0	0	1	0	0	0
	0	0	*	*	*	*	*	*	*	*	*	*
Ring	0	0	0	0	0	0	0	0	1	0	0	0
	0	0	0	0	0	0	0	0	0	0	0	0
	0	0	*	*	*	*	*	*	*	*	*	*

however, that on the average an adult male baboon weighs about twice as much as an adult female, and this discrepancy in body weight alone may account for the observed pattern of adult male-adult female dominance relationships.

Table IX summarizes the outcomes of decided agonistic bouts between adult females and males of the subadult class or younger. The males are listed by age class and within each age class by dominance rank. These younger males defeated adult females in 677 agonistic bouts (table IX A) and were in turn defeated by adult females in 203 bouts (table IX B).

Based on table IX the following ontogeny of dominance relationship between maturing males and adult females is hypothesized: As an infant, a male will be consistently subordinate to all adult females with whom he interacts, though he will probably be involved in agonistic bouts with only a few adult females. During the J1 age period (1–2 years of age), a male will continue to maintain relatively consistent relationships with all adult females and will be dominant to a few low-ranking adult females.

During the J2 age period (2–4 years of age), a male will become dominant to an increasing number of adult females, but will remain consistently subordinate to at least a few adult females. Dominance relationships between a J2 male and adult females will be marked by numerous periods of inconsistency, particularly when compared to the relationship between adult females and males either older or younger than the J2 class. As noted above, all undecided agonistic bouts recorded in the focal female samples involved J2 male participants. As a subadult (4–6 years of age), a male will be consistently dominant to all but one or two adult females. Finally, as shown in table VIII, adult males are consistently dominant to all adult females.

Peer Group Dominance Relationships

Table X lists all observed outcomes of decided agonistic bouts between all individuals of the J2 class or younger as recorded in the *ad libitum* sample records. Within those juvenile classes for which some information was available, males were dominant to females of the same age. Thus J2 female Fem was subordinate to the four J2 males, and both J1 females Vee and Gin were subordinate to J1 males Swat and Spot. In fact, Swat, the largest J1 male, was dominant also to J2 female Fem. It should be pointed out that this pattern of male-female dominance relationship might be accounted for either by sex differences in agonistic behavior patterns, by sex-related differences in body-size and strength, or by maternal effects on dominance relationships, as discussed below.

Table X. Outcome of agonistic bouts between individuals of the J2 class or younger, *ad libitum* sample record. Names in parentheses indicate the presumed mother of the juvenile. Data on bouts between juvenile females are the same as those presented in table VI. Other conventions as in table VI

Winner	Loser															
	J2♂♂				J2♀, Fem	J1♂♂			J1♀♀		I2♂♂		I2♀♀		I1♂, I-Ring	I1♀, I-Fluff
	Stiff	Stu	Red	Russ		Swat	Major	Spot	Vee	Gin	Kub	Dogo	Bell	Mindi		
J2♂♂																
Stiff	–	14	27	21	15	10	0	13	4	13	6	1	0	0	0	0
Stu	0	–	16	11	23	9	2	11	7	12	1	1	0	0	0	0
Red	0	3	–	45	22	21	1	40	16	14	9	5	0	0	0	0
Russ	0	0	0	–	8	21	0	21	8	17	5	5	0	0	0	0
J2♀																
Fem	0	0	0	0	–	0	0	2	5	36	1	1	0	0	0	0
J1♂♂																
Swat	0	0	0	0	15	–	0	34	18	16	18	13	0	0	0	0
Major (Skinny)	0	0	0	0	0	0	–	2	0	0	0	0	0	0	0	0
Spot (Alto)	0	0	0	0	0	0	1	–	16	27	2	12	0	0	0	0
J1♀♀																
Vee (Fluff)	0	0	0	0	1	0	0	0	–	10	0	2	0	0	0	0
Gin (Ring)	0	0	0	0	0	0	0	0	0	–	0	0	0	0	0	0
I2♂♂																
Kub (Mom)	0	0	0	0	1	0	0	0	1	2	–	3	0	0	0	0
Dogo (Preg)	0	0	0	0	0	0	0	0	0	0	0	–	0	0	0	0
I2♀♀																
Bell (TT)	0	0	0	0	0	0	0	0	0	0	0	0	–	0	0	0
Mindi (Scar)	0	0	0	0	0	0	0	0	0	0	0	0	0	–	0	0
I1♂																
I-Ring (Ring)	0	0	0	0	0	0	0	0	0	0	0	0	0	0	–	0
I1♀																
I-Fluff (Fluff)	0	0	0	0	0	0	0	0	0	0	0	0	0	0	0	–

Maternal Effects on Dominance Rank of Offspring

Each of the juveniles listed in table X from Major through Infant-of-Fluff was seen to suckle exclusively from one adult female; it is not known whether these females were still lactating. The female from whom each juvenile suckled was presumed to be its mother and is listed in table X also. Within those pairs of juveniles of like age and sex for whom mothers could be identified in this way, the dominance relationship between the offspring was the same as that between the putative mothers:

Kub, the I2 male offspring of Mom, defeated Dogo, the I2 male offspring of Preg, in 3 agonistic bouts; Dogo was not observed to defeat Kub in any bouts. Vee, the J1 female offspring of Fluff, defeated Gin, the J1 female offspring of Ring, in 10 agonistic bouts; Gin was not observed to defeat Vee in any bout. Spot, the J1 male offspring of Alto, defeated Major, the J1 male offspring of Skinny, in one bout near the beginning of the study, though Major defeated Spot in two bouts thereafter. Major, together with female Skinny (possibly his mother), disappeared from the group during the second month of the study.

Thus, the dominance relationship between this last pair of juveniles underwent one change during the short portion of the study that Major was with the group. However, at the time of his disappearance, Major was dominant to Spot. My sample contained inadequate data to analyze the ontogeny of dominance relationships between immature male offspring of low ranking adult females and immature female offspring of high ranking adult females; their pattern of dominance relationship remains to be determined.

Male-Male Dominance Relationships

Table XI summarizes the outcomes of decided bouts of agonistic behavior among all males as recorded in the *ad libitum* sample record during the study period. In general, males of the older age classes were consistently dominant to males of the younger classes.

Within the nonadult classes of males, dominance relationships within all but two pairs of males were consistent over the entire study period. The dominance relationship between Spot and Major was reviewed above. The dominance relationship between J2 males Stu and Red was as follows:

From the beginning of the study on 1 August 1971 through 8 October 1971, Red was observed to defeat Stu in 3 agonistic bouts; Stu was not observed to defeat Red in any bouts. Between 9 October 1971 and 3 November 1971, no agonistic bouts, decided or undecided, were observed between Stu and Red. However, on 4 November 1971, Stu defeated Red in an agonistic bout and did so in 15 more bouts through the end of the study. Red was not observed to defeat Stu from 9 October 1971 through the end of the study.

Table XI. Outcome of decided agonistic bouts between males, *ad libitum* sample record. Data on bouts between juvenile males are the same as those presented in table X. Other conventions as in table VI

| Winner | Loser | | | | | | | | | Sub-adult ♂♂ | | J2♂♂ | | | | J1♂♂ | | | I2♂♂ | | I1♂ |
| | Adult ♂♂ | | | | | | | | | | | | | | | | | | | | |
	BJ	Peter	Ivan	Crest	Stubby	Dutch	Sinister	Max	Cowlick	Ben	Even	Stiff	Stu	Red	Russ	Swat	Major	Spot	Kub	Dogo	I-Ring
Adult ♂♂																					
BJ	–	67	61	41	82	131	35	77	54	71	32	17	15	9	11	5	0	1	0	4	0
Peter	0	–	38	30	79	48	22	32	32	33	19	9	2	6	6	1	0	2	0	0	0
Ivan	11	48	–	12	103	185	55	117	93	94	42	35	21	21	32	6	0	27	1	1	0
Crest	0	0	0	–	34	50	11	18	14	10	2	5	8	4	4	2	0	4	1	1	0
Stubby	10	30	91	4	–	84	38	97	59	139	73	24	26	28	19	11	1	4	5	0	0
Dutch	0	0	0	14	0	–	31	48	58	23	23	12	6	10	12	6	0	8	0	1	0
Sinister	0	0	0	2	0	0	–	38	19	17	16	10	8	9	12	4	0	3	2	3	0
Max	0	0	0	0	0	0	0	–	34	52	30	14	16	14	13	8	0	17	3	3	0
Cowlick	0	0	0	2	0	18	24	20	–	27	15	2	5	6	5	2	0	1	0	0	0
Subadult ♂♂																					
Ben	0	0	0	0	0	0	0	0	0	–	71	21	26	26	26	7	0	7	3	1	0
Even	0	0	0	0	0	0	0	0	0	0	–	41	47	31	47	16	1	16	3	4	0
J2♂♂																					
Stiff	0	0	0	0	0	0	0	0	0	0	0	–	14	27	21	10	0	13	6	1	0
Stu	0	0	0	0	0	0	0	0	0	0	0	0	–	16	11	9	2	11	1	1	0
Red	0	0	0	0	0	0	0	0	0	0	0	0	3	–	45	21	1	40	9	5	0
Russ	0	0	0	0	0	0	0	0	0	0	0	0	0	0	–	21	0	21	5	5	0
J1♂♂																					
Swat	0	0	0	0	0	0	0	0	0	0	0	0	0	0	0	–	0	34	18	13	0
Major	0	0	0	0	0	0	0	0	0	0	0	0	0	0	0	0	–	2	0	0	0
Spot	0	0	0	0	0	0	0	0	0	0	0	0	0	0	0	0	1	–	2	12	0
I2♂♂																					
Kub	0	0	0	0	0	0	0	0	0	0	0	0	0	0	0	0	0	0	–	3	0
Dogo	0	0	0	0	0	0	0	0	0	0	0	0	0	0	0	0	0	0	0	–	0
I1♂																					
I-Ring	0	0	0	0	0	0	0	0	0	0	0	0	0	0	0	0	0	0	0	0	0

These two pairs of juvenile males in which a period of inconsistent dominance relationship occurred constituted 3% of all juvenile male pairs present in the group at the beginning of the study (N = 66 pairs).

Dominance Relationships Among Adult Males

The outcome of all decided agonistic bouts between adult males as recorded in *ad libitum* samples are shown in table XI also. Data from the focal male samples are presented in table XV A and will be discussed below. Although there are numerous reversals (figures below the main diagonal) in this table, these reversals were not produced by a random series of wins and losses within each pair of adult males. Instead, the dominance relationship within each pair of adult males was consistent over varying periods of time, with only brief periods of inconsistent dominance relationship separating the periods of consistent relationship. The majority of undecided agonistic bouts occurred during these brief periods of inconsistency or in close proximity to such periods.

Periods of inconsistent dominance relationship within a pair of adult males, as inconsistency is defined in this work, usually resulted in a change in the dominance ordering in that pair (i.e. the new winner in a pair also won the next bout between the two), but this was not always the case (table XIV). When a change in the dominance ordering of adult males did occur, it was not a general reshuffling of all ranks among all males, but was restricted to one or two pairs of males who occupied adjacent dominance ranks. Table XI obscures not only the short-term consistency of adult male dominance relationships, but also the fact that not all adult males were with Alto's group for the entire study period (table III).

Tables XII and XIII are a replotting of the data on outcome of decided agonistic bouts for certain pairs of adult males, as recorded in *ad libitum* (table XI) and focal-male (table XV A) samples, and these tables provide a more clear demonstration of the short-term consistency of dominance relationships among adult males in the study group. For clarity, several short-term changes in the adult male composition of the study group have been deleted from these tables (cf. table III). Data on the outcomes of decided agonistic bouts between the pairs of adult males shown in tables XII and XIII were divided into a series of matrices, in such a way as to minimize reversals (entries below the main diagonal), and in most of these matrices, no outcomes appear below the main diagonal. In those few matrices that contain reversals, the reversal resulted either from a decided agonistic bout within a pair of males that had already entered a period of inconsistent

Table XII. Partitioned agonistic bout outcome data for adult males BJ, Stubby, Ivan, Peter, Crest (1972 only). See text for explanation

(a)
1–7 August 1971

	Ivan	Stubby	BJ	Peter
Ivan	–	3	1	1
Stubby	0	–	1	1
BJ	0	0	–	1
Peter	0	0	0	–

BJ left group on 5 Aug. 1971.

(b)
8–27 August 1971

	Stubby	Ivan	Peter
Stubby	–	15	2
Ivan	0	–	2
Peter	0	0	–

8–11 Aug. 1971: Ivan–Stubby inconsistent, 2 undecided bouts.

(c)
28 August–21 September 1971

	Stubby	Peter	Ivan
Stubby	–	6	25
Peter	0	–	4
Ivan	0	0	–

28 Aug.–8 Sept. 1971: Ivan–Peter inconsistent, 8 undecided bouts.
After 8 Sept. 1971: Peter–Ivan, 2 undecided bouts. Stubby–Ivan, 2 undecided bouts.

(d)
22–26 September 1971

	Stubby	Ivan	Peter
Stubby	–	5	2
Ivan	0	–	2
Peter	0	0	–

22 Sept. 1971: Peter–Ivan inconsistent.

After 22 Sept. 1971: Peter–Ivan, 1 undecided bout.

(e)
27–29 September 1971

	Stubby	Peter	Ivan
Stubby	–	0	4
Peter	0	–	3
Ivan	0	0	–

27–28 Sept. 1971: Ivan–Peter inconsistent.

(f)
30 September–6 October 1971

	Stubby	Ivan	Peter
Stubby	0	2	6
Ivan	0	–	5
Peter	0	0	–

30 Sept.–4 Oct. 1971: Ivan–Peter inconsistent.

Table XII (continued)

(g)
7–16 October 1971

	Stubby	Peter	Ivan
Stubby	–	2	6
Peter	0	–	13
Ivan	0	0	13

7 Oct. 1971: Ivan–Peter inconsistent.

(h)
17 October–5 November 1971

	Stubby	Ivan	BJ	Peter
Stubby	–	12	11	6
Ivan	0	–	7	6
BJ	0	0	–	7
Peter	0	0	0	–

17–28 Oct. 1971: Ivan–Peter inconsistent, 2 undecided bouts. 28 Oct. 1971: BJ rejoins group. Other dates: BJ–Peter, 1 undecided bout. BJ–Stubby, 3 undecided bouts, BJ-Ivan, 1 undecided bout.

(i)
6–9 November 1971

	Stubby	Ivan	BJ	Peter
Stubby	–	2	1	3
Ivan	0	–	7	4
BJ	2	0	–	1
Peter	0	0	0	–

6–9 Nov. 1971: BJ–Stubby inconsistent, 1 undecided bout. BJ–Ivan remained consistent.

(j)
10 November 1971–1 January 1972

	BJ	Stubby	Ivan	Peter
BJ	–	22	24	9
Stubby	0	–	26	3
Ivan	0	0	–	15
Peter	0	0	0	–

10–15 Nov. 1971: BJ–Stubby inconsistent, 3 undecided bouts. BJ–Ivan inconsistent, 3 undecided bouts.

(k)
2 January–2 May 1972

	BJ	Ivan	Peter	Stubby
BJ	–	39	46	43
Ivan	0	–	31	103
Peter	0	0	–	57
Stubby	0	0	0	–

2–3 Jan. 1972: Stubby–Ivan inconsistent. Stubby–Peter inconsistent. No undecided bouts.

(l)
3–11 May 1972

	BJ	Peter	Ivan	Stubby
BJ	–	3	0	3
Peter	0	–	0	3
Ivan	0	0	–	9
Stubby	0	0	0	–

3–11 May 1972: Ivan–Peter inconsistent, 1 undecided bout.
11 May 1972: Ivan–Stubby inconsistent, 1 undecided bout.
11 May 1972: Ivan–Stubby inconsistent, see below also.
Other dates: Ivan–BJ, 1 undecided bout.

Table XII (continued)

(m)			
12–24 May 1972			

	BJ	Peter	Stubby	Ivan
BJ	–	8	10	6
Peter	0	–	11	17
Stubby	0	0	–	2
Ivan	0	0	0	–

12–13 May 1972: Stubby–Ivan inconsistent, 1 undecided bout.
Other dates: Peter–Ivan 2 undecided bouts. Ivan–BJ 1 undecided bout.

(n)			
25–30 May 1972			

	BJ	Peter	Ivan	Stubby
BJ	–	3	2	1
Peter	0	–	4	6
Ivan	0	0	–	1
Stubby	0	0	0	–

25–30 May 1972: Ivan–Stubby inconsistent (Ivan defeated Stubby in 1 decided bout, but failed to win the subsequent bout).

(o)			
31 May–3 June 1972			

	BJ	Peter	Stubby	Ivan
BJ	–	1	1	2
Peter	0	–	0	1
Stubby	0	0	–	2
Ivan	0	0	0	–

31 May–3 June 1972: Ivan–Stubby consistent, 2 undecided bouts.

(p)			
4 June–7 July 1972			

	BJ	Peter	Ivan	Stubby
BJ	–	16	7	14
Peter	0	–	5	7
Ivan	0	0	–	9
Stubby	0	0	0	–

4–6 June 1972: Stubby–Ivan inconsistent, 1 undecided bout. 25 June 1972: Ivan disappeared from group.

(q)			
8 July–3 September 1972			

	BJ	Peter	Crest	Stubby
BJ	–	6	29	10
Peter	0	–	29	11
Crest	0	0	–	39
Stubby	0	0	0	–

8 July 1972: Crest rejoins group.

Table XIII. Partitioned agonistic bout outcome data for adult males Dutch, Cowlick, Sinister, Max, and Crest (1971 only). See text for explanation

(a)
1 August–25 October 1971

	Cow.	D.	S.	M.
Cowlick	–	18	10	14
Dutch	0	–	6	7
Sinister	0	0	–	10
Max	0	0	0	–

(b)
26–30 October 1971

	D.	Cow.	S.	M.
Dutch	–	10	0	6
Cowlick	0	–	0	1
Sinister	0	0	–	1
Max	0	0	0	–

26 Oct. 1971: Cowlick–Dutch inconsistent.

(c)
31 October–3 November 1971

	D.	Cow.	Cr.	S.	M.
Dutch	–	4	8	0	6
Cowlick	0	–	2	0	0
Crest	0	0	–	2	1
Sinister	0	0	0	–	0
Max	0	0	0	0	–

(d)
4–22 November 1971

	D.	Cow.	S.	Cr.	M.
Dutch	–	3	3	6	0
Cowlick	0	–	1	0	1
Sinister	0	0	–	2	3
Crest	0	0	0	–	1
Max	0	0	0	0	–

4–12 Nov. 1971: Crest–Sinister inconsistent. 22 Nov. 1971: Crest leaves group.

(e)
23–26 November 1971

	D.	S.	Cow.	M.
Dutch	–	0	8	0
Sinister	0	–	5	0
Cowlick	0	0	–	4
Max	0	0	0	–

23–25 Nov. 1971: Cowlick–Sinister inconsistent.
Other dates: Dutch–Cowlick, 1 undecided bout.

(f)
27 November 1971–27 February 1972

	D.	S.	M.	Cow.
Dutch	–	11	12	30
Sinister	0	–	8	14
Max	0	0	–	19
Cowlick	0	0	0	–

27 Nov.–2 Dec. 1971: Cowlick–Max inconsistent.

Table XIII (continued)

(g)
28 February–24 May 1972

	D.	S.	M.
Dutch	–	6	12
Sinister	0	–	1
Max	0	0	–

28 Feb. 1972: Cowlick leaves group.

(h)
25 May–12 June 1972

	D.	S.	M.	Cow.
Dutch	–	3	2	2
Sinister	0	–	5	0
Max	0	0	–	4
Cowlick	0	22	0	–

25 May 1972: Cowlick rejoins group. Cowlick, Max, Sinister in triangular dominance relationship at rank 7.

(i)
13–24 June 1972

	D.	S.	M.
Dutch	–	1	3
Sinister	0	–	2
Max	0	0	–

13 June 1972: Cowlick leaves group.

(j)
25 June 1972

	D.	S.	M.	Cow.
Dutch	–	0	2	0
Sinister	0	–	0	0
Max	0	0	–	0
Cowlick	0	3	0	–

25 June 1972: Cowlick rejoins the group. Cowlick, Sinister, Max in triangular dominance relationship at rank 6.

(k)
26–27 June 1972

	D.	S.	M.	Cow.
Dutch	–	3	4	1
Sinister	0	–	3	0
Max	0	0	–	0
Cowlick	0	0	0	–

26–27 June 1972: Cowlick–Sinister inconsistent. See next matrix also.

(l)
28–30 June 1972

	D.	S.	M.	Cow.
Dutch	–	0	0	1
Sinister	0	–	0	4
Max	1	0	–	1
Cowlick	0	0	0	–

28–29 June 1972: Sinister–Cowlick inconsistent, 1 undecided bout. Four decided bouts (Sinister winner) on 30 June 1972.
28–30 June 1972: Dutch–Max inconsistent. Max defeated Dutch in a decided bout on 30 June 1972, but failed to win the subsequent bout.

Table XIII (continued)

(m)
1 July–3 September 1972

	D.	S.	M.	Cow.
Dutch	–	2	8	7
Sinister	0	–	9	4
Max	0	0	–	13
Cowlick	0	0	0	–

dominance relationship (tables XII i, XIII l), or from decided agonistic bouts among three adult males who were in a period of triangular (i.e. consistent but not transitive) dominance relationship (tables XIII h, j). Additionally, for each time period shown in these tables, the duration of all inconsistent dominance relationships within pairs of adult males and the number of undecided agonistic bouts within each pair are listed below the matrix for that time period.

Tables XII and XIII demonstrate that dominance relationships within pairs of adult males occupying adjacent dominance ranks were consistent over varying periods of time, and that changes occurred frequently in the dominance ordering within such pairs. These tables also substantiate that periods of inconsistent dominance relationship within pairs of adult males were relatively brief and that the undecided agonistic bouts within a pair of adult males most frequently occurred in close proximity to a period of inconsistent dominance relationship between those males. However, the method of outcome data presentation used in these tables suffers from a number of deficits, not the least of which is that the large number of separate tables needed for each pair of adult males makes the total pattern of changes in the dominance ordering of adult males difficult to comprehend. Additionally, it may be argued that the actual partitioning of these data was arbitrary, or at least that the method of partitioning was not explicitly stated.

Figure 9 is yet another method of presenting dominance relationships among adult males in the study group, particularly emphasizing periods of inconsistency within pairs of males and changes in the dominance ordering of males. This figure is based upon all outcomes of decided agonistic bouts obtained in both focal male (table XV A) and *ad libitum* (table XI) samples of behavior, and as in tables XII and XIII, 20 short-term changes in the adult male composition of the group have been omitted for clarity (table III).

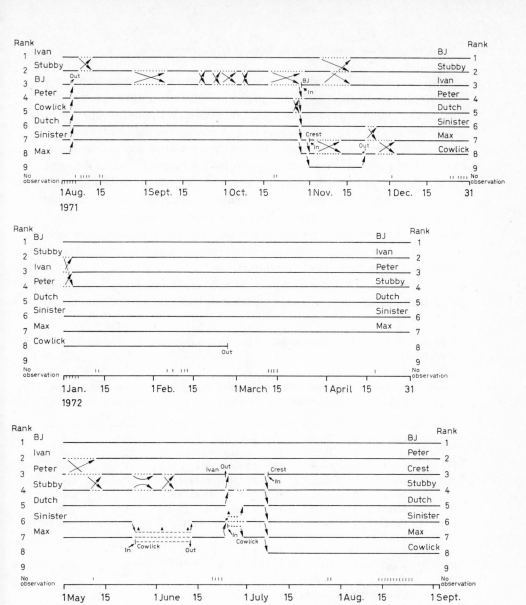

Fig. 9. Consistency of adult male dominance relationships. A solid line indicates that a male's dominance relationship with all other males was consistent, a dotted line indicates that a male's dominance relationship with at least one other male was inconsistent, and a dashed line with a small triangle above it indicates periods of consistent, but nontransitive dominance relationship among three or more males. Days on which no behavioral observations were actually completed are indicated at the bottom of this figure. 20 short-term changes (less than one week) in the adult male composition of the group have been deleted for clarity (cf. table III).

In figure 9, a solid line indicates a period when an adult male's dominance relationships with all other adult males were consistent, while a dotted line indicates a period that an adult male's dominance relationship with one or more other adult males was inconsistent. A dashed line indicates the special case of consistent, but non-transitive dominance relationships referred to as triangularity and resulting in three males having assigned to them the same rank number. In figure 9, a day on which no behavior observations were made is indicated at the bottom of the figure and the dominance relationship within pairs of adult males on such days has been inferred from the relationship on the immediately preceding day of observation. Observations were not made on an average of three days per month, though only one absence was greater than four days in total duration.

In fact, figure 9 somewhat visually overemphasizes periods of inconsistent dominance relationship among adult males: A dotted line was entered

Table XIV. Date and duration of periods of inconsistent dominance relationship for pairs of adult males

Number	Former winner and rank		Former loser and rank		Dates of inconsistent period	Duration, days (n = 104)	Rank change
1	Ivan	1	Stubby	2	8–11 Aug. 1971	4	yes
2	Ivan	2	Peter	3	28 Aug.–8 Sept. 1971	12	yes
3	Peter	2	Ivan	3	22 Sept. 1971	1	yes
4	Ivan	2	Peter	3	27–28 Sept. 1971	2	yes
5	Peter	2	Ivan	3	30 Sept.–4 Oct. 1971	5	yes
6	Ivan	2	Peter	3	7 Oct. 1971	1	yes
7	Peter	2	Ivan	3	17–28 Oct. 1971	12	yes
8	Cowlick	4	Dutch	5	26 Oct. 1971	1	yes
9	Stubby	1	BJ	3	6–15 Nov. 1971	10	yes
10	Ivan	2	BJ	3	10–15 Nov. 1971	6	yes
11	Crest	7	Sinister	8	4–12 Nov. 1971	9	yes
12	Cowlick	6	Sinister	7	23–25 Nov. 1971	3	yes
13	Cowlick	7	Max	8	27 Nov.–2 Dec. 1971	6	yes
14	Stubby	2	Ivan	3	2–3 Jan. 1972	2	yes
15	Stubby	2	Peter	4	2–3 Jan. 1972	2	yes
16	Ivan	2	Peter	3	3–11 May 1972	9	yes
17	Ivan	2	Stubby	4	11–13 May 1972	3	yes
18	Stubby	3	Ivan	4	25–30 May 1972	6	no
19	Stubby	3	Ivan	4	4–6 June 1972	3	yes
20	Sinister	6	Cowlick	6	26–29 June 1972	4	yes
21	Dutch	4	Max	6	28–30 June 1972	3	no

in this figure when an adult male's dominance relationship with one or more other adult males was inconsistent, even though his relationship with each of the remaining adult males in the group may have continued to be consistent. Nevertheless, the short-term consistency of adult male dominance relationships, the brief duration of periods of inconsistent dominance relationship, and the restriction of such inconsistent periods and of rank changes in general to adult males occupying adjacent dominance ranks are clearly demonstrated in figure 9. Additionally, one can directly determine from figure 9 the number of days that a male's dominance relationship with every other male was consistent or inconsistent (table XIV) as well as the total duration of rank occupancy for each male, or time-in-rank (see below).

Focal Male Samples

Altogether, 1,086 bouts of agonistic behavior among adult and subadult males of Alto's group were recorded in 222.10 h of focal male samples taken from 7 March 1972 through 3 September 1972. Twenty-seven (2.5%) of these agonistic bouts were undecided by the outcome criteria described above; the outcomes of the remaining 1059 agonistic bouts observed in the focal male samples are presented in table XVA. Table XVB presents comparable data from the *ad libitum* samples for the same time period. Males Crest and Ivan were never simultaneously in Alto's group during this period (table III) and therefore these tables provide information on the dominance relationship in 54 pairs of adult or subadult males. Of these 54 pairs, the dominance relationship in 50 remained consistent over this entire six-month period.

The focal animal samples provided an unbiased sample of outcomes of agonistic bouts, whereas the randomness of the sample of outcomes provided by *ad libitum* samples has been questioned [ALTMANN, 1974]. However, as was the case with the *ad libitum* and focal data for adult females (tables VI, VII), the outcome data for agonistic bouts within pairs of adult and subadult males recorded in the *ad libitum* samples (table XVB) and the comparable data from focal male samples (table XV A) show close agreement as to the consistency and asymmetry of dominance relationships among adult males. Thus, if the *ad libitum* samples did provide a biased sample of bout outcomes, the magnitude of this bias must have been small relative to the asymmetry of paired relationships in adult males and adult females. Similarly, during periods of consistent dominance relationship within the four pairs of males that underwent a period of inconsistency during the time period covered in table XV, the *ad libitum* samples and focal male samples show close agreement (table XVI).

Table XV. Comparison of results of *ad libitum* and focal male samples. *A* Outcome of all decided agonistic bouts between adult and subadult males as recorded in focal male samples. Focal male samples were taken only from 7 March 1972 through 3 September 1972. *B* Outcome of all decided agonistic bouts between adult and subadult males as recorded in the *ad libitum* sample record for the comparable dates. Other conventions as in tables VI and VII

Winner	Loser										
	Adult ♂♂									Subadult ♂♂	
	BJ	Peter	Ivan	Crest	Stubby	Dutch	Sinister	Max	Cowlick	Ben	Even
A. Focal male data											
Adult ♂♂											
BJ	–	26	22	26	35	47	17	36	40	33	13
Peter	0	–	18	22	45	17	10	8	10	14	7
Ivan	0	14	–	*	52	56	22	45	22	30	22
Crest	0	0	0	–	34	50	9	16	14	9	1
Stubby	0	0	4	0	–	21	11	31	13	43	31
Dutch	0	0	0	0	0	–	15	21	14	11	14
Sinister	0	0	0	0	0	0	–	18	7	10	8
Max	0	0	0	0	0	1	0	–	19	21	12
Cowlick	0	0	0	0	0	0	13	0	–	13	10
Subadult ♂♂											
Ben	0	0	0	0	0	0	0	0	0	–	33
Even	0	0	0	0	0	0	0	0	0	0	–
B. Ad libitum data											
Adult ♂♂											
BJ	–	18	7	3	14	12	11	7	1	20	12
Peter	0	–	9	7	16	9	8	7	5	12	8
Ivan	0	9	–	*	22	25	22	9	6	23	6
Crest	0	0	0	–	5	12	2	5	5	5	3
Stubby	0	0	0	0	–	10	9	17	21	27	28
Dutch	0	0	0	0	0	–	4	14	8	9	5
Sinister	0	0	0	0	0	0	–	8	8	4	5
Max	0	0	0	0	0	1	0	–	3	17	11
Cowlick	0	0	0	0	0	0	3	0	–	4	5
Subadult ♂♂											
Ben	0	0	0	0	0	0	0	0	0	–	22
Even	0	0	0	0	0	0	0	0	0	0	–

Table XVI. Comparison of results of *ad libitum* and focal male samples for four pairs of males that underwent a period of inconsistent dominance relationships, 7 March 1972 through 3 September 1972

Adult ♂ pair	Consistent period	Domi- nant ♂	*Ad libitum* sample results			Focal male sample results		
			ad libitum sample min	bouts won by domi- nant ♂	bouts lost by domi- nant ♂	focal sample min	bouts won by domi- nant ♂	bouts lost by domi- nant ♂
1. Ivan-	1 Mar–2 May 1972	Ivan	27,777	14	0	1,070.35	9	0
Peter	12 May–24 June 1972	Peter	20,181	18	0	780.00	9	0
2. Ivan-	1 Mar–10 May 1972	Ivan	31,853	41	0	1,406.51	22	0
Stubby	14–24 May 1972	Stubby	5,790	2	0	330.00	0	0
	31 May–3 June 1972	Stubby	2,403	2	0	90.00	0	0
	7–24 June 1972	Ivan	5,720	11	0	210.00	0	0
3. Dutch-	1 Mar.–27 June 1972	Dutch	53,833	13	0	1,691.09	15	0
Max	1 July–3 Sept. 1972	Dutch	17,006	8	0	654.37	0	0
4. Cowlick-	25 May–12 June 1972	Cowlick	9,396	13	0	264.14	3	0
Sinister	30 June–3 Sept. 1972	Sinister	17,589	7	0	708.30	8	0

Protocols of Adult Male Dominance Relationships

In the following protocols, changes in the dominance relationships and dominance rank of three adult males are reviewed. Adult male BJ was first ranking adult male during most of the study period, while adult males Dutch and Cowlick were fifth or middle ranking and eighth or lowest ranking, respectively, during most of the study period. Thus, the history of the dominance relationships of these three males provides a representative sample of adult male dominance relationships in the study group.

Adult Male BJ

When this study began, BJ was third ranking adult male in Alto's group. On 2–4 August 1971, BJ was in consort with an estrous female, Lulu, though this consortship was abruptly terminated on the afternoon of 4 August 1971 when BJ received a long, deep wound on his left hindleg. (The fight was not observed.) Adult male Ivan, then the first ranking individual in the group, subsequently consorted with Lulu. On the morning of 5 August 1971 BJ limped away from the main portion of Alto's group and lay on the ground in the shade of a *Salvadora persica* bush; his leg wound dripped fluid and his locomotion was greatly impaired. During most of that morning, BJ trailed along a quarter of a mile behind the group, though in the afternoon, he walked away from the group and spent the night far removed from the group. BJ was a solitary male until late August, 1971, when he was seen traveling with BTF group (table I), but by 3 October 1971, BJ was again solitary.

On the morning of 26 October 1971, a lone adult male was observed about one quarter mile north of Alto's group, and this male was later identified as BJ, both by his face and tail carriage and by the prominent scar on his left hindleg. Adult male Stubby was then the first ranking individual in the group, and Stubby left the group and walked directly toward BJ. A lengthy chase ensued and continued for the next two hours. During the chase, Stubby repeatedly forced BJ into the terminal branches of a yellow barked acacia tree *(Acacia xanthophloea)* and BJ continually directed submissive behaviors toward Stubby. When the chase ended, both males were out of visual contact with Alto's group. Stubby returned to the group within the next half hour, but BJ was not seen again that day.

However, on the evening of 27 October 1971, BJ came walking casually into the middle of Alto's group as they sat feeding beneath their sleeping trees; BJ's arrival was greeted by a volley of trilled 'cohesion' grunting from the other group members. Within the next half hour, BJ presented to, and grimaced at, adult males Stubby and Ivan, the first and second ranking males, respectively, while, in turn, adult males Peter and Max, the third and eight ranking males, respectively, presented to, and grimaced at, BJ. There was no overt aggression between Stubby and BJ, though Stubby did approach and displace BJ several times before the group ascended the sleeping trees for the night. BJ spent that night with the group and in the days that followed was third ranking adult male in the group, the same rank that he had occupied prior to his departure in August, 1971.

From 6 November through 15 November 1971, the dominance relationship between BJ and Stubby was inconsistent. BJ defeated Stubby in an agonistic bout on 9 November 1971 and on 12 November 1971 Stubby defeated BJ in an agonistic bout. A total of four undecided agonistic bouts between Stubby and BJ was also recorded during this period of inconsistency, while three undecided agonistic bouts preceded the inconsistent period. In these undecided bouts, Stubby strutted directly toward BJ who responded by leaning aside and avoiding eye contact. However, Stubby responded by halting his approach about 5 ft from BJ, leaning aside from BJ and avoiding eye contact with BJ; then Stubby turned sharply and trotted away. The dominance relationship between BJ and second ranking adult male Ivan became inconsistent on 10 November 1971 and three undecided agonistic bouts of a similar pattern were observed between BJ and Ivan during their period of inconsistent relationship.

Finally, on 16 November 1971, BJ defeated both Stubby and Ivan in several agonistic bouts and did so repeatedly from then until the end of the study in September, 1972. Thus, BJ remained consistently dominant to both Stubby and Ivan after 16 November 1971, though two undecided agonistic bouts between BJ and Ivan were observed prior to the end of the study.

Adult Male Dutch

Dutch was sixth ranking adult male in Alto's group at the beginning of the study. As a result of BJ's departure from the group in August, 1971 (see above), Dutch moved up one position in the adult male dominance order, to fifth ranking adult male. Through 25 October 1971, Dutch maintained consistent dominance relationships with all other adult males in the group, and in particular was consistently defeated in agonistic bouts by Cowlick, the fourth ranking adult male. However, on or about 21 October 1971, Cowlick received a severe wound to his right hindfoot (see below). By 26 October 1971, this wound had apparently become infected and Cowlick appeared physically weakened. On 27 October 1971, Dutch defeated Cowlick in an agonistic bout, and Dutch henceforth consistently defeated Cowlick in agonistic bouts through the end of the study, at least during those periods that Cowlick was in Alto's group (table III and below). BJ returned to Alto's group one day later, on 28 October (see above). The net result of BJ's return and of the change of rank between Cowlick and Dutch was that Dutch remained fifth-ranking adult male in Alto's group.

Dutch continued to occupy fifth rank until 25 June 1972 when adult male Ivan disappeared from Alto's group (table III); Dutch then moved to fourth rank. However, on the morning of 30 June 1972, Dutch was the first individual in Alto's group to descend from the sleeping trees. Shortly after Dutch's descent, adult male Max, then sixth ranking in the group, came from the trees and walked directly toward Dutch. Max stood in front of Dutch and yawned widely, exposing his canine teeth in profile. Max also made exaggerated chewing motions with his jaws, scratched his neck and shoulders vigorously, and directed other harassing behaviors toward Dutch. Dutch avoided eye contact with Max then suddenly jumped up, put his tail vertically into the air, and grimaced widely at Max. Next, Dutch turned and grabbed a nearby black infant (I1), put the infant into the ventral carry position, and ran away from Max; Max pursued Dutch only briefly. On two other occasions that morning Max walked to within 50 ft of Dutch who casually moved closer to a nearby black infant, but otherwise Dutch did not direct any clearly submissive behaviors toward Max. No further agonistic bouts between Dutch and Max were observed until 1 July 1972 when Dutch successfully defeated Max in an agonistic bout, and Dutch consistently did so through the end of the study. Finally, on 8 July 1972, adult male Crest rejoined Alto's group (table III) as third ranking adult male and hence Dutch had returned to fifth ranking adult male in Alto's group by the end of the study.

Adult Male Cowlick

Cowlick was the fifth ranking adult male in Alto's group when the study began, but soon moved to fourth rank with BJ's departure from the group (table III and above). On or about 21 October 1971, Cowlick received a deep puncture wound to his right hindfoot and began limping badly. On 27 October and subsequently, Dutch defeated Cowlick in agonistic bouts (see above). BJ rejoined Alto's group near this time and the net result of Cowlick's defeat by Dutch and BJ's return was that Cowlick fell to sixth ranking adult male. On 30 October through 2 November 1971, Cowlick followed Alto's group away from the sleeping groves in the mornings, but he had such great difficulty in walking that he lagged far behind the group. By midmorning each day, Cowlick returned to the sleeping groves by himself and remained there alone until Alto's group returned in the evening. Cowlick's wound eventually healed and through 23 November 1971, he maintained consistent dominance relationships with all other adult males in the study group. However, on 26 November 1971, adult male Sinister defeated Cowlick in an agonistic bout and on 3 December 1971, adult male Max did likewise. Thus, by 3 November 1971, Cowlick had fallen to eighth ranking adult male in Alto's group.

On the morning of 28 February 1972, Cowlick, adult female Ring, and juvenile (J1) female Gin, moved away from the sleeping groves with another nearby group (High Tail's group) rather than with Alto's group; a more complete description of this departure is given in chapter V. (Near the end of 1972, these two groups amalgamated.) Ring and Gin returned to Alto's group on 3 March 1972 (table III), but Cowlick remained with High Tail's group. However, on those occasions that High Tail's group foraged or slept near Alto's group, Cowlick often sat between the two social groups and was frequently groomed by female Gin and juvenile (J2) female Fem, who I presume to be Gin's sister. My wife, SUE ANN HAUSFATER, who was studying High Tail's group during this period, reported that Cowlick often left that group completely for periods of half a day or longer. Cowlick's return to High Tail's group after such a brief absence, or his proximity to Alto's group during intergroup contact, occasioned no special greeting behaviors from members of either group.

In contrast, on 26 May 1972, Cowlick returned to Alto's group after a long absence and was greeted by a loud volley of trilled grunting from the group. Within the next half hour, Cowlick was defeated in agonistic bouts by adult males Ivan and Max, then fourth and seventh ranking adult males, respectively. However, also within the hour Cowlick successfully defeated adult male Sinister, then sixth ranking adult male, in several agonistic bouts, even though Sinister had previously been dominant to Cowlick. Since Sinister remained dominant to Max, and Max was dominant to Cowlick, and Cowlick was now dominant to Sinister, all three of these males were assigned to rank number seven and a period of triangular dominance relationship was begun.

On 12 June 1972, Cowlick again left Alto's group and rejoined High Tail's group. On 26 June 1972, Cowlick returned to Alto's group and was still dominant to Sinister, but subordinate to Max, though the departure of adult male Ivan in the interim (table III) meant that Sinister, Max, and Cowlick were each assigned to rank number six, rather than rank number seven. Next, on 30 June 1972, Sinister defeated Cowlick in several agonistic bouts and Cowlick fell to rank number seven. Cowlick was consistently defeated by Sinister and Max in agonistic bouts through the end of the study, but with Crest's return to the group in early July, 1972, Cowlick moved to eighth ranking adult male in Alto's group where he remained through the end of the study.

Comparison of Adult Male and
Adult Female Dominance
Relationships in the Study Group

Changes in the rank of an individual and the duration of his occupancy of each rank may have important consequences for his reproduction and survival. For example, the priority-of-access model of mating behavior, to be tested in chapter IV, hypothesizes that the dominance rank of an adult male is an important determinant of his reproductive success. Additionally, the frequency of changes in the rank ordering of a class of animals and differences in their patterns of rank occupancy provide a useful quantitative means of analyzing the stability of dominance relationships within classes of animals as well as a means of making comparisons between classes. In the remainder of this chapter, several components of the stability of dominance relationships within a social group will be analyzed, including frequency of changes in rank, magnitude of rank changes, rank occupancy patterns, and duration of periods of inconsistent dominance relationship.

Frequency of Rank Changes by
Adult Males and Adult Females

The protocols, given above, of the dominance relationships of adult males BJ, Dutch, and Cowlick, together with figure 9, demonstrate that changes in dominance rank within the study group were either the result of demographic changes, such as one individual entering or leaving the group, or the result of actual changes in the dominance relationships within one or two pairs of individuals already in the group. These latter changes will be referred to as 'agonistically induced' rank changes, and the former will be referred to as 'demographically induced' rank changes. Agonistically induced rank changes affected pairs of individuals, and during the 14-month study period no individuals ever moved up or down more than two ranks on a single day (table XIV). Demographically induced rank changes affected a much more variable number of individuals, depending on the rank of the migrant. Thus, when third ranking adult male BJ left Alto's group in August, 1971, all five adult males that remained in the group and that previously ranked below BJ were each automatically assigned the next higher rank. In contrast, when the lowest ranking adult male, Cowlick, left Alto's group in February, 1972, all of the seven adult males that remained in the group had ranked higher than Cowlick before his departure and thus did not undergo any changes in dominance rank.

Demographic Changes in the Number of
Adult Males and Adult Females in the Study Group

During this study, all demographic changes in the number of adult males and adult females in the study group resulted from either death, emigration, or immigration; no juveniles matured into the adult classes while the group was under study. It is assumed that all changes in the adult male or adult female composition of the group that occurred during the 400-day study period were recorded. The present study group, Alto's group, contained a mean of 7.5 adult males during the study period, and table III lists 30 changes (15 emigrations or deaths and 15 immigrations) in the adult male composition of the study group (of which 10 are shown in fig. 9). Therefore, a demographic change in the number of adult males in Alto's group occurred on the average once every 13.3 study days (400 study days/30 changes), and each of these changes could have potentially affected the ranking of adult males in the study group, depending on the rank of the migrant male. Another way to look at these potential demographically induced changes in the ranking of adult males is in terms of adult male-days of study: One adult male-day of study is defined as one adult male present in the group on one study day. A total of 3,029 adult male-days of study on Alto's group were completed. Thus, a demographic change that could have resulted in a change in the ranking of adult males in the study group occurred on the average once every 101.0 adult male-days of study (3,029 adult male-days of study/ 30 changes).

Of the 30 changes in the adult male composition of the study group listed in table III, 20 resulted from an individual male leaving and then subsequently reentering the group after only a few days absence, usually less than one week. During one of these short-term changes, the adult male was not with another social group, but apparently was solitary. The remaining 10 changes in the adult male composition of Alto's group resulted from more durative (greater than one week) group changes by adult males, and it is these durative changes in the adult male composition of the group that are shown in figure 9. One of these durative group changes by an adult male occurred on the average once every 40.0 study days (400 study days/10 durative changes) or, in other terms, once every 302.9 adult male-days of study (3,029 adult male-days of study/10 durative changes), and each such demographic change could have potentially changed the ranking of adult males in the group.

Alto's group contained a mean of 10.2 adult females during the study period. In contrast to the adult males, only seven changes (5 emigrations or deaths and 2 immigrations) in the adult female composition of Alto's group occurred during the study period (table III). Thus, a demographic change in the number of adult females in Alto's group occurred on the average once every 57.1 study days (400 study days/7 changes) or, in other terms, once every 583.6 adult female-days of study (4,085 adult female-days of study/ 7 changes). Finally, it should be noted that a comparison of the frequency of actual or potential demographically induced changes in the ranking of adult males and adult females in the study group is more a statement about class- and rank-specific rates of immigration, emigration, and death for adult males and adult females, than a statement about the inherent instability of dominance relationships *per se* within these classes. The relationship, if any, between the drastic ecological changes taking place in the Amboseli study area and the high frequency of group changes found in this study are not yet known. Nevertheless, it is important to note that many changes in the rank ordering of adult males and adult females resulted merely from natural demographic processes.

Agonistically Induced Rank Changes

Although several short breaks in observation did occur (fig. 9, bottom lines), it seems likely that all agonistically induced rank changes during the 400-day study period were recorded. At least on our return from short trips away from the Reserve, the dominance order among adult males in the study group was found in all cases to be the same as it had been prior to our departure. All agonistically induced rank changes by adult males were preceded by a short period of inconsistent dominance relationship (as indicated by a dotted line in fig. 9), though not all periods of inconsistent dominance relationship between a pair of males resulted in a dominance rank change. Of the 21 separate periods of inconsistent dominance relationship within pairs of adult males in the study group, 19 resulted in a change in the dominance relationship within the pair, and hence an agonistically induced rank change (table XIV). Thus, an agonistically induced rank change among adult males in the study group occurred on the average once every 21.0 study days (400 study days/19 changes). This figure of 21.0 days between changes is, of course, specific to the study group. Another way to look at these data is in terms of dyad-days of study: A dyad-day of study is defined as one pair of individuals present in the group on one study day, and a total of 10,025 adult male dyad-days of study were accumulated in the 400-day study period. This gives a mean of 25.1 adult male pairs, or dyads, present in the group during the study period and an agonistically induced rank change within a pair of adult males once every 527.6 adult male dyad-days of study (10,025 adult male dyad-days of study/19 agonistically induced changes).

In contrast, no agonistically induced changes in rank number occurred within pairs of adult females during the 400-day study period. In Alto's group, there was a mean of 47.2 adult female pairs, or dyads, present during the study period; and a total of 18,884 adult female dyad-days of study were accumulated. Thus, an agonistically induced change in dominance ranking within a pair of adult females would be expected at some interval greater than 18,884 adult female dyad-days of study. In sum, the adult female dominance relationships in the study group were far more stable than those of adult males regardless of whether one judges stability by the frequency of demographically induced or agonistically induced rank changes.

Magnitude of Rank Number Changes in Adult Males

Table XVII is a matrix tabulation of all changes in rank among adult males as recorded in figure 9 and, as in that figure, the effect of 20 short-term changes (less than one week) in the adult male composition of the group have been ignored. The upper number in a given cell of table XVII shows the total number of times that an adult male moved from the rank indicated by the row of that cell to the rank indicated by the column of that cell, regardless of whether the change was agonistically or demographically induced. The second entry in each cell gives only the frequency of agonistically induced rank changes for that cell. The row labeled 'immigrated' indicates the rank assigned to adult males who joined the study group, while the 'emigrated' column shows the rank assigned to adult males prior to their departure from the group. These data, of course, can be used to calculate the probability that during a rank change, an adult male occupying any given rank will move to any other rank, or leave the group entirely.

As noted above, each demographically induced rank change affected a variable number of individuals, depending on the rank of the migrant individual. Of necessity, however, each adult male affected by a demographically induced change moved up, or down, only one rank number (fig. 9; see also table III for deleted demographic changes). In contrast, there was no logical restriction on the extent of the rank changes induced agonistically. Although agonistically induced rank changes were defined as occurring within pairs of individuals, it was still possible for one adult male simultaneously to change his dominance relationship in two of the pairs of which he was a member and thus move, say, from rank 3 to rank 1 as did adult male BJ in November 1971.

Interestingly, table XVII shows only 4 (7.0%) cases of an adult male actually moving up or down by more than one rank in a single rank change (N = 57 rank changes). In all four of these changes, the adult male moved up or down by two ranks, and these will be referred to as 'two-step' rank changes. Three of these two-step rank changes were in fact agonistically induced, while the fourth two-step rank change among adult males resulted from the combined effects of an agonistically induced rank change and a demographically induced rank change that occurred almost simultaneously (Peter and Ivan, 28–29 October 1972, fig. 9). The three agonistically induced two-step rank changes constituted only 8.1% of all agonistically induced rank changes among adult males (N = 37 agonistically induced rank

Table XVII. Rank change matrix. The first number in each cell indicates the frequency that adult males changed from the rank indicated by the row of the cell to the rank indicated by the column of the cell. The second entry in each cell is the number of such changes that were agonistically induced. See text for further explanation

Old rank	New rank									Emigrated
	1	2	3	4	5	6	7	8	9	
	–	2	0	0	0	0	0	0	0	0
1	–	2	0	0	0	0	0	0	0	0
	1	–	6	3	0	0	0	0	0	0
2	1	–	6	3	0	0	0	0	0	0
	1	8	–	2	0	0	0	0	0	2
3	1	8	–	1	0	0	0	0	0	0
	0	0	5	–	3	0	0	0	0	0
4	0	0	3	–	1	0	0	0	0	0
	0	0	0	3	–	2	0	0	0	0
5	0	0	0	1	–	0	0	0	0	0
	0	0	0	0	2	–	5	0	0	0
6	0	0	0	0	1	–	3	0	0	0
	0	0	0	0	0	4	–	5	0	1
7	0	0	0	0	0	2	–	2	0	0
	0	0	0	0	0	0	3	–	1	2
8	0	0	0	0	0	0	2	–	0	0
	0	0	0	0	0	0	0	1	–	0
9	0	0	0	0	0	0	0	0	–	0
	0	0	2	0	0	1	2	0	0	–
Immigrated	0	0	0	0	0	0	0	0	0	–

changes). Thus, agonistically induced rank changes in adult males were primarily exchanges of rank between two males that occupied adjacent ranks in the dominance order, and in general, only rarely did a rank change in adult males, whatever its causation, result in an individual adult male moving up or down more than two ranks.

Patterns of Rank Occupancy by Adult Males

Figure 10 is a tabulation of the duration of periods of continuous occupancy by adult males for each dominance rank and for all dominance ranks combined. For the purposes of this analysis, a rank is considered 'occupied' by an adult male only when that male's dominance relationships with *all* other adult males in the study group were consistent. Thus, each period of continuous rank occupancy tabulated in figure 10 corresponds to one of the solid or dashed lines in figure 9 and as in that figure the effects

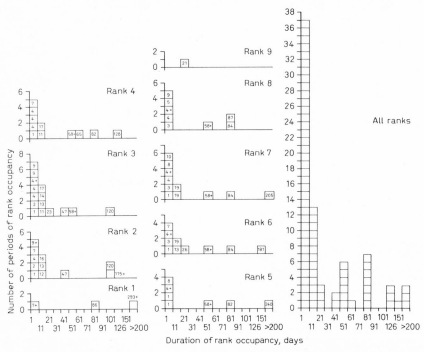

Fig. 10. Rank occupancy durations. The figure shows the frequency distribution of rank occupancy durations taken from figure 9 tabulated by rank of male. Numbers inside of squares indicate the precise duration of each period of rank occupancy. Each period of rank occupancy in this figure corresponds to one period of consistent relationship (a solid or dashed line) shown in figure 9.

of 20 short-term changes in the adult male composition of the group have been deleted (cf. table III). Additionally, during the few brief periods of triangular dominance relationship, the three adult males involved were each assigned the same dominance rank, and as a result of this joint assignment of males to a single rank, for example rank 7, the next higher and next lower ranks, rank 6 and 8 respectively, were left temporarily vacant.

The duration of rank occupancy data for some periods are, of course, incomplete. In particular, the exact duration of rank occupancy is unknown for eight rank-occupancy periods (one per adult male) that were already in progress when the study began and for eight that were in progress at the termination of the study. Duration of rank occupancy data derived from these 16 truncated intervals has been included in figure 10, but has been marked with a plus sign in the figure.

The duration of rank occupancy data for all ranks combined, when the truncated intervals are deleted, suggests a smooth unimodal distribution, sharply skewed to the left (figure 10). The modal rank occupancy duration interval is 1–10 days, while the median duration of rank occupancy fell between 11 and 12 days. A larger sample of such duration of rank occupancy data may reveal a bimodal distribution, with the second mode falling in the 81- to 90-day interval (figure 10).

For ranks 1–9, the mean number of distinct periods of rank occupancy by adult males was 8.5 periods; in other words, the adult male occupant of a particular dominance rank changed on the average 8.5 times during the study period. It should be pointed out that if a specific individual adult male lost, then subsequently regained, a given rank, he alone would account for two distinct periods of occupancy at the given rank. Thus, the statement that there was on the average 8.5 distinct periods of occupancy by adult males for each dominance rank does not mean that on the average 8.5 *different* adult males occupied each rank. In fact, most ranks were occupied more than once by some adult male (fig. 9).

Figure 11 presents the duration of periods of rank occupancy ordered by the identity of the adult male occupant rather than by the rank occupied. These data, then, show the time-in-rank for individual adult males who were with Alto's group for any length of time during the study period as well as the mean rank occupied by that male while he was in the group. The mean number of ranks occupied per adult male was 3.6 and this figure does not change if adult males who were not with the group for the entire study period are excluded from the calculation. Thus, whether the rank occupancy data are ordered by rank of occupant or by identity of occupant, it is clear that

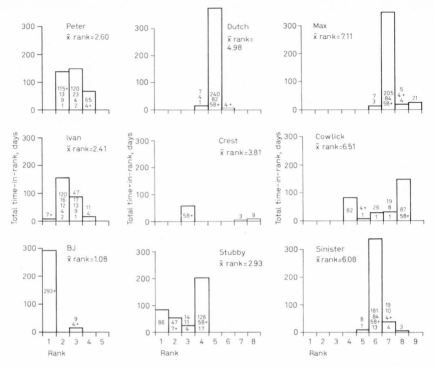

Fig. 11. Total time-in-rank for individual adult males. The figure shows the frequency distribution of rank occupancy durations, or time-in-rank, tabulated by identity of male. Same data set as shown in figures 9 and 10. Conventions as in figure 10.

adult males in the study group changed their rank frequently during the study period and, conversely, that no dominance rank was occupied exclusively by one particular adult male.

Duration of Inconsistent Dominance Relationships in Adult Males

Figure 12 is a tabulation of the duration of periods of inconsistent dominance relationship for adult males ordered by rank of male prior to the onset of the inconsistent period. Thus, each such period plotted in figure 12 corresponds to one of the dotted lines in figure 9. As in figure 10, the distribution of duration of periods of inconsistent dominance relationship is unimodal and sharply skewed to the left; however, the modal duration in figure 12 is 1–3 days and the median duration of periods of inconsistent dominance relationship was between 3 and 4 days. Thus, although both periods of consistent and inconsistent dominance relationship were usually

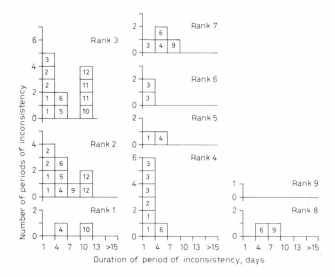

Fig. 12. Duration of inconsistent dominance relationships, showing the frequency distribution of durations of period of inconsistent dominance relationship in adult males. Numbers inside of bars show the precise duration of each period of inconsistency. Each period of inconsistency in this figure corresponds to one of the dotted lines shown in figure 9.

of short duration, periods of inconsistent dominance relationship generally were more brief than periods of consistent dominance relationship. The mean number of periods of inconsistent dominance relationship for all ranks combined was 5.0, though the adult males occupying ranks 2 through 4 accounted for 70% (N = 40) of all periods of inconsistent dominance relationship.

Sex Differences in Inconsistency of Dominance Relationship

Another way of expressing sex differences in the stability of dominance relationships is as follows: Table XIV shows that pairs of adult males were inconsistent in their dominance relationship for a total of 104 days, or 104 adult male dyad-days of inconsistent dominance relationship. The exact duration of these periods of inconsistent dominance relationship within pairs of adult males was discussed above. However, these 104 adult male dyad-days of inconsistent dominance relationship equal approximately 1% of all adult male dyad-days of study ($N = 10,025$). In contrast, the one period of inconsistent dominance relationship within a pair of adult females, Preg and Scar (table VI), lasted for only one day, or for very much less than 0.01% of all adult female dyad-days of study ($N = 18,884$).

Comparison with Other Baboon Studies

In contrast to the present results, HALL and DEVORE [1965], HALL [1962] and DEVORE [1965] reported that dominance relationships among adult males and among adult females in their study groups of anubis *(Papio anubis)* and chacma *(P. chacma)* baboons were not adequately described by the concept of a linear dominance order. In particular, these authors concluded that the rank of an adult male as an individual was different from his rank when acting in coalition with certain other 'central' males, and the dominance rank of adult females was said to fluctuate depending on their relationship with adult males in the group and on their reproductive condition. It should be emphasized that the above statements are the conclusions of HALL and DEVORE; data supporting these conclusions have not been published and thus cannot be evaluated in their own right. In fact, DEVORE's few published data on adult male and adult female dominance relationships [HALL and DEVORE, 1965] contradict the above conclusions: they are consistent with a linear dominance order [ALTMANN, 1965]. The hypothesis of HALL and DEVORE that the dominance rank of an adult female changed as a result of sexual cycling will be tested in chapter V.

STOLTZ and SAAYMAN [1970] concluded that dominance relationships among adult males in their study group of chacma baboons fit the same pattern described by HALL and DEVORE [1965]. STOLTZ and SAAYMAN also concluded that dominance relationships among adult females in their study group were either subtly expressed or entirely absent; however, no data were presented to support either this or the previous conclusion. In contrast, ROWELL [1966a] concluded that all agonistic behavior among *adult males* in forest-living groups of anubis baboons in Uganda was rare and that the dominance relationships among adult males were not overtly expressed.

ROWELL [1966b] also has published data showing consistent linear dominance relationships among adult females in a captive group of anubis baboons, while KUMMER [1956] reported similar findings in his study of a captive group of hamadryas baboons *(P. hamadryas)*. Both KUMMER and

Rowell used the outcome of agonistic bouts as their criteria of dominance, and both workers found that the dominance order of females based on agonistic behavior was predictive of many other social relationships. In contrast to the results of his zoo study, Kummer [1968] concluded that there was little evidence of 'social bonds' among adult females in one-male units of hamadryas in the wild, though a description of agonistic behavior among adult females was published.

Concerning the dominance relationships of adult males in troops of hamadryas baboons, Kummer [1968, p. 153] stated:

'Dominance in the usual restricted meaning, as the ability of one animal to displace another from an incentive, can be assumed to exist among hamadryas males, but its manifestations in troop life are minimal if compared with anubis baboons or rhesus monkeys.'

Kummer further concluded that the ability of an adult male to gain attention and compliance of other males in the choice of travel routes was probably a more important characteristic of the male than dominance rank.

In many ways, there is substantial agreement among these published reports on wild baboon populations. Thus, Hall and DeVore, Stoltz and Saayman, and Kummer, all emphasize the importance of relations among adult males for the organization of the group, and describe the behavior of females as being strongly dependent on relationships with adult males. The dominance rank of an adult male as an individual is minimized compared to his ability to enlist other adult males in coalitions, or to influence their choice of travel routes. Yet these authors have, for the most part, failed to publish data that support their conclusions. Furthermore, I believe that each of these studies has been handicapped to a greater or lesser degree by lack of identifiable individuals, particularly identifiable adult females, short study duration, and by lack of systematic sampling of dominance relationships.

Altmann [1974] has pointed out that behavior studies that rely entirely on *ad libitum* or other nonsystematic samples run the danger of recording only the most conspicuous social interactions. This is particularly true in observational studies of baboons where the large size of the adult male compared to the adult female or juvenile results in the observer's attention more frequently being drawn to adult males than to other classes of animals. In particular, noisy squabbles and chases between adult males are far more attention attracting than are the less violent agonistic bouts between adult females or juveniles. Thus, I suspect that most previous field studies of baboons have been biased against the behavior of females and juveniles; in other words, the near universal agreement on the importance of adult male relationships in the above field studies is the result of consistent observational bias against the behavior of females and juveniles rather than an inherent aspect of group organization. The bias against inconspicuous behaviors or smaller individuals is often eliminated in studies of confined groups, since all animals are constantly in view of the observer and often only one fully adult male is present. Thus, it is not surprising that studies of confined groups of baboons [ROWELL, 1966b; KUMMER, 1956] have concluded that consistent social relationships among adult females are both present and of importance for understanding group organization.

Although it is intended to extend the present observations over some longer time period, the strategy of focusing on one single study group, identifying all group members, and sampling their behavior systematically, is a wise practice even in short-term research. Thus, while DEVORE [1965] concluded: 'the adult male dominance hierarchy constitutes the most stable and rigidly maintained series of dominance relationships in the baboon troop,' the present data indicate that the dominance relationships among adult females are far more stable than those of adult males. Similarly, KUMMER [1968] has emphasized the importance of adult males at all levels of social organization in hamadryas baboons. However, the fact that peer-group dominance relationships of juveniles, at least in the present study, were largely a reflection of dominance relationships among adult females is potentially a more important aspect of group organization than any aspect of adult male behavior. The long-term study of rhesus monkeys *(Macaca mulatta)* on Cayo Santiago by SADE [1972] has provided many more examples of the importance of maternal geneologies in group organization and individual behavior. As SADE has pointed out, long-term and systematic studies of groups of identifiable individuals reveal many aspects of social organization and behavior that no amount of short-term study of individuals of unknown history would ever reveal. Thus, a longitudinal study of baboons stands to make a significant contribution to our understanding of primate social behavior, organization, and ecology.

Summary

1. The behaviors of aggression and submission in baboons have been reviewed and criteria for determining the outcome of agonistic bouts described. Dominance, as used in this work, is a statement about the consistency of outcome of successive decided dyadic agonistic bouts within pairs of

identifiable individuals. Each individual was assigned a dominance rank number equal to the number of the individuals in the group minus the number of other individuals to whom he was consistently dominant. Multi-individual agonistic bouts were infrequent, and the dominance relationships within pairs of individuals showed no modification as a result of participation by either member of the pair in a multi-individual coalition against the other member.

2. Dominance relationships between females of all ages showed strong consistency through time, and all adult females could be arranged in a single linear dominance order throughout the study period. Adult males also could be arranged into a single linear dominance order on most study days, though brief periods of inconsistent dominance relationships within pairs of adult males were recorded. Dominance relationships within the nonadult classes of males showed strong consistency through time and nonadult males could also be linearly ordered throughout most of the study period. In general, among both males and females, individuals of the older age classes were consistently dominant to individuals of the younger age classes.

3. All adult males were consistently dominant to all adult females. Within other age classes, males were consistently dominant to females also. It is hypothesized that as a male matures, he becomes consistently dominant to an increasingly large number of females of all ages, including females older than himself.

4. The dominance relationships of juveniles of like age and sex whose mothers were presumed known correlated with the dominance relationships of the presumed mothers.

5. Comparison of the sample of outcomes of agonistic bouts recorded in *ad libitum* samples with the unbiased sample of outcomes recorded in focal animal samples indicated that the pattern of dominance relationships summarized in 2–4 above was not the result of observational bias toward certain bout outcomes.

6. Changes in the rank numbering of individuals were divided into demographically induced rank changes and agonistically induced rank changes. A potential demographically induced change in the rank numbering of adult males occurred, on the average, once every 13.3 study days, while a comparable potential change in the ranking of adult females occurred, on the average, only once every 57.1 study days. An agonistically induced change in the ranking of adult males occurred, on the average, once every 21.0 study days, while no agonistically induced changes in the ranking of adult females occurred in the entire 400-day study period. Also, pairs of

adult males were inconsistent in their dominance relationship for a far larger proportion of study days than were pairs of adult females, though in both sexes, pairs of individuals were inconsistent in their dominance relationship for only a small proportion of all study days.

7. Previous studies of baboons have been reviewed and it has been suggested that the use of nonsystematic observational techniques and lack of identifiable individuals has produced a consistent observational bias against the behavior of adult females and juveniles. Thus, it is probable that the importance of adult male relationships for group organization has been overemphasized in most previous studies of baboons. The exceptional stability of adult female dominance relationships, and the influence of a mother's rank on the rank of her offspring, may be potentially more important in group organization than any aspect of adult male behavior. The need for systematic longitudinal studies on groups of identifiable individuals is apparent.

III. The Menstrual Cycle and Estrous Behavior of Baboons

Introduction

The length of the menstrual cycle in baboons and the proportion of the cycle for which the female is fertile are important parameters of the priority-of-access model of mating behavior to be tested in chapter IV. This chapter summarizes available laboratory and field data on the menstrual cycle and estrous behavior of baboons. Data on menstrual cycle length for females in the study group will be presented and compared with data from other baboon populations. Laboratory data on the timing of ovulation and fertile matings within the menstrual cycle will be reviewed and several perineal and behavioral indications of ovulation will be discussed. The chapter will conclude with a brief analysis of rates of social interaction of estrous females and thus provides the necessary contextual information for understanding mating patterns in baboons (chapter IV).

The Menstrual Cycle

Cycle Length

The menstrual cycle of chacma baboons *(Papio ursinus)* has been analyzed in great detail by GILLMAN and GILBERT [1946] and cyclical changes in the external genitalia described [GILLMAN, 1935]. More recently, detailed analyses of menstrual cycle length, cyclical changes in vaginal cytology and endometrial histology, and cyclical changes in ovarian hormone levels in anubis *(P. anubis)* and yellow baboons *(P. cynocephalus)* have been published by several authors [HENDRICKX, 1971; MACLENNAN *et al.*, 1971 a, b; STEVENS *et al.*, 1970]. The baboons living in the Amboseli study area are classified as *Papio cynocephalus*, yellow baboons, by HILL [1967].

ZUCKERMAN [1937] presented data on 32 cycles of two yellow baboons maintained in captivity at the London Zoological Gardens and calculated a mean cycle length of 33.3 days (table XVIII, part B). The mean duration between successive menstruations of 96 yellow baboons maintained at the Southwest Foundation in San Antonio, Texas, was 33.2 days (table XVIII, part B) [HENDRICKX and KRAEMER, 1969, 1971]. Several other authors have published cycle length data for baboons, but have either failed to identify the species studied or have combined data from two or more species. This is an important consideration since consistent differences between *Papio cynocephalus*, yellow baboons, and *P. anubis*, anubis baboons, in cycle length and swelling morphology have been reported [HENDRICKX, 1971; MACLENNAN and WYNN, 1971]. Available information on the duration of menstrual cycles in other baboon species has been summarized recently by HENDRICKX and KRAEMER [1971] and will not be repeated here.

The perineal condition of every adult and J2 female in the study group was noted on every day of observation. The size of each female's sexual skin swelling was scored on a scale of 0, swelling absent, to 20, the maximum swelling any female ever achieved (fig. 13), and daily sexual skin swelling charts for every female in the group are available in HAUSFATER [1974]. In practice, the largest swelling size most females reached was 10 or 12, though female Lulu had a swelling size of 20 on nearly all cycles. Although the scoring of sexual skin size was somewhat a subjective process, the day of onset of rapid deturgescence of the swollen sexual skin was unambiguously identifiable.

Six females in Alto's group (Alto, Fluff, Oval, Lulu, Judy, and Ring) each cycled between 3 and 14 times during the study period (table XVIII). These females provided both behavioral and cycle length data. Additionally, five females (Scar, TT, Mom, New, and Skinny) underwent only a portion of a menstrual cycle during the study and therefore did not contribute any cycle length data. Some behavioral observations were made on these latter females.

The length of the menstrual cycles of females in the study group has been determined in two ways (table XVIII). The mean interval between successive onsets of rapid deturgescence of the sexual skin for 34 pairs of successive cycles recorded in this study was 32.5 days (table XVIII, part A2). The mean interval between successive onsets of externally visible menses was 31.4 days (N = 26 pairs of successive cycles) (table XVIII, part A1). For any given set of data, mean interdeturgescence interval, as calculated here (table XVIII, part A2), and mean inter-menstruation interval, as calculated here and by other workers (table XVIII, parts A1 and B), will yield the same estimates of mean cycle length but, in general, different estimates of cycle length variance. For logistic reasons, the onset of menstruation of females in the study group was often missed, but every day of onset of rapid deturgescence was recorded. Thus, estimation of mean cycle length from inter-deturgescence interval allowed more complete utilization of the available data. This method of cycle length estimation also would prove useful in studies of those species

A, B

C

Fig. 13. Sexual skin swellings. A Female Alto with estrus 0 swelling. Alto was not cycling when photographed. B Female Ring with estrus 5 swelling, about halfway to maximum turgescence. Photographed on cycle day D-13. C Female Oval with estrus 10 swelling peak, turgescence, ejaculate in vulva (and on left leg), and wound on right lateral lobe of swelling. Photographed on cycle day D-1.

Table XVIII. Menstrual cycle length (days), giving the mean cycle length for females in the study group calculated by two methods described in text and data from comparable baboon populations. Raw cycle length data for females in study group are given in HAUSFATER [1974]

Data source	Mean	SE of mean	SD	Range	N
A. Alto's group					
1. Intermenstruation interval					
Lulu	30.1	0.3	1.1	29–39	11
Judy	29.7	0.8	2.4	28–35	10
Oval	40.5	5.5	7.8	35–46	2
Alto	–	–	–	–	–
Fluff	35.0	12.0	17.0	23–47	2
Ring	38.0	–	–	38	1
All females	31.4	1.0	5.3	23–47	26
All females except Lulu	32.4	1.8	6.8	34–57	15
2. Interdeturgescence Interval					
Lulu	30.0	0.3	1.1	28–31	13
Judy	28.5	0.5	1.6	26–30	10
Oval	40.7	5.0	9.9	31–52	4
Alto	39.0	8.0	11.3	31–47	2
Fluff	38.0	5.6	9.6	27–45	3
Ring	37.5	1.5	2.1	36–39	2
All females	32.5	1.1	6.5	26–52	34
All females except Lulu	34.0	1.7	7.9	26–52	21
3. Deturgescent phase					
Lulu	14.9	0.3	1.0	13–16	12
Judy	12.7	0.3	0.8	12–14	10
Oval	13.3	0.7	1.1	12–14	3
Alto	15.0	–	–	15	1
Fluff	13.0	1.1	2.0	11–15	3
Ring	15.5	0.5	0.7	15–16	2
All females	13.9	0.3	1.5	11–16	31
All females except Lulu	13.3	0.3	1.4	11–16	19
4. Turgescent phase					
Lulu	15.0	0.4	1.2	13–17	12
Judy	16.7	0.8	2.7	12–19	12
Oval	23.0	3.2	6.4	17–32	4
Alto	32.0	–	–	32	1
Fluff	24.3	6.5	11.2	12–34	3
Ring	22.0	1.0	1.4	21–33	2
All females	18.5	1.0	5.7	12–34	35
All females except Lulu	20.3	1.3	6.3	12–34	23

Table XVIII (continued)

Data source	Mean	SE of mean	SD	Range	N
B. Other populations					
1. London Zoological Gardens [ZUCKERMAN, 1937] intermenstruation interval, 2 females	33.3	0.7	3.5	25–41	32
2. Southwest Foundation for Research and Education [HENDRICKX and KRAEMER, 1969] intermenstruation interval, 32 females	33.2	–	3.7	19–43	96

that do not exhibit externally visible menstruation, but do show cyclical changes in the external genitalia, e.g. labial swelling.

Lulu

Lulu cycled regularly for the entire 14-month study period and exhibited one of the largest sexual skin swellings observed in any wild baboon. However, swellings as large as Lulu's are common in captive baboons held in isolation [ROWELL, 1970]. Possibly Lulu's large swelling and failure to become pregnant are indicative of a pathological abnormality. Therefore all cycle length estimates in table XVIII and behavioral data in tables XIX and XX have been given both including and excluding Lulu's cycles from the sample.

Lulu showed no obvious aberrations in estrous behavior or in cycle-length, nor did she appear either more or less attractive to males than females with smaller swellings. At least two other females in this population cycled for the entire study period also. The cycle length estimates and behavioral observations that include Lulu are the most useful for an observer entering a new study population, since every population probably has a few females who exhibit putatively abnormal swellings.

Hormonal Basis of Sexual Skin Changes

The onset of first day of rapid deturgescence of the swollen sexual skin divides the baboon menstrual cycle into two distinct phases, a turgescent phase and a deturgescent phase. The turgescent phase extends from the onset of menstruation up to, but not including, the first day of rapid deturgescence. The deturgescence phase begins on the first day of rapid deturgescence and continues up to the subsequent menstruation. The turgescent phase of the baboon menstrual cycle corresponds to the period of follicular activity in the ovarian cycle, while the deturgescent phase corresponds to the luteal portion of the ovarian cycle [GILLMAN and GILBERT, 1946; STEVENS *et al.*, 1970]. The mean length of the turgescent phase of cycles analyzed in this study was 18.5 days, while mean length of the deturgescent phase was 13.9 days (table XVIII). Variation in the duration of the turgescent phase of the menstrual cycle accounted for most of the variability in total cycle length.

GILLMAN and his co-workers [GILLMAN, 1940a, 1942; GILLMAN and STEIN, 1941; GILLMAN and GILBERT, 1946] have demonstrated that turgescence of the baboon sexual skin is estrogen-induced and that deturgescence of the sexual skin is stimulated by a substance with progesterone-like activity. Estrogen production increases throughout the turgescent phase of the cycle, and increased estrogen production continues in the last week of the turgescent phase even though changes in the size of the sexual skin may be slight [GILLMAN and GILBERT, 1946].

Partial deturgescence of the sexual skin may be induced by withdrawal of estrogen during turgescence, but complete deturgescence will occur only with the active production of progesterone by the corpus luteum. Very small amounts of progesterone will produce complete deturgescence followed by menstruation even when high concentrations of estrogen still are present in the female's system [GILLMAN, 1940b; GILLMAN and STEIN, 1941].

Recent experiments by MICHAEL *et al.* [1966] and MICHAEL and WELLEGALLA [1968] have shown that cyclical changes in behavior can be induced in ovariectomized female rhesus monkeys *(Macaca mulatta)* by estrogen and progesterone replacement therapy, and the cyclical changes in social behavior of female baboons probably have the same underlying hormonal mechanism as cyclical changes in the sexual skin.

Cycle Day Notation

For convenience, the following shorthand notation for specifying menstrual cycle day has been used in this work. Each cycle day was classified with respect to the day of onset of rapid deturgescence of the swollen sexual skin, or 'D-day'. The first day preceding the day of onset of deturgescence was labeled 'D-1', the second preceding day 'D-2', etc. Cycle days which followed the day of onset of rapid deturgescence were labeled 'D+1',

'D + 2', 'D + 3', etc., for the first, second, third, etc., day following the day of onset of deturgescence (D-day).

The term 'estrus' as used in this work refers to the seven-day period of the baboon menstrual cycle immediately preceding the day of deturgescence of the swollen sexual skin, namely cycle days D-7 through D-1 inclusive. Since, as will be shown below, ovulation and maximal sexual receptivity occur within this seven-day period, this usage of the term 'estrus' is consistent with the traditional use of the term by mammalogists [COCKRUM, 1962]. Although in some primate species there may be difficulty in distinguishing estrous periods from other nonfertile periods of sexual attractiveness [LOY, 1970], this is not a problem in baboons where the sexual skin provides a good indication of cycle state.

Ovulation and Fertile Matings

GILLMAN and GILBERT [1946] performed several laparotomies on chacma baboons *(Papio ursinus)* and concluded that ovulation and fertile matings in their laboratory colony occurred two or three days prior to rapid deturgescence of the sexual skin, i.e. cycle days D-2 or D-3. HENDRICKX [personal commun.] stated that serial laparotomy and analysis of hormonal levels in several female baboons *(P. anubis* or *P. cynocephalus)* confirmed that ovulation occurs most often on or around cycle day D-3. HENDRICKX further concluded that while ovulation may occur later than D-3, it is extremely unlikely to occur earlier.

In a series of controlled mating experiments on baboons at the Southwest Foundation for Research and Education, HENDRICKX and KRAEMER [1971] found that single periods of mating on cycle day D-3 had the highest probability of conception, and that single periods of mating on cycle day D-3 alone resulted in nearly the same proportion of conceptions (41%) as repeated matings over the entire week preceding rapid deturgescence, i. e. cycle days D-7 through D-1. Although single periods of mating on all cycle days from D-7 through D-4 had a nonzero conception rate (less than 10%) also, HENDRICKX [personal commun.] suggests that these high conception rates may be an artifact of prolonged sperm viability under laboratory conditions, small sample size or other factors. Evidence from several other mammalian species indicates that the viability of sperm and egg is generally limited to less than 24 h [HAMMOND and ASDELL, 1926; HARTMAN, 1924] and that the optimal period for fertilization is the first few hours after ovulation [ALLEN *et al.*, 1928; HUNTER and DZIUK, 1968; OVERSTREET and ADAMS, 1971]. Thus, while the possibility that fertile matings in baboons can occur earlier than cycle day D-3 cannot definitely be excluded, it seems most probable that under natural conditions fertile matings occur almost exclusively on cycle day D-3 or one day later.

Perineal Indicators of Ovulation

Several investigators [GILLMAN and GILBERT, 1946; MACLENNAN and WYNN, 1971; HENDRICKX and KRAEMER, 1971] have noted that a slight transient decrease in the size of the swollen sexual skin often occurred on cycle day D-2 or D-3 in baboons. A similar phenomenon was observed frequently during the present study [HAUSFATER, 1974]. GILLMAN and GILBERT hypothesized that this transient depression in sexual skin swelling size coincided with ovulation. However, while MACLENNAN and WYNN [1971] also frequently observed a transient decline in swelling size in their study animals, they concluded that this decline did not correlate positively with ovulation. Furthermore, MACLENNAN and WYNN were often able to detect the transient decline in swelling size as early as cycle day D-4.

HENDRICKX and KRAEMER [1971] have reviewed their attempts to find an external indication of ovulation in baboons. They conclude that achievement of maximum turgescence is the best available external predictor of proximity to ovulation. Knowledge of days since last menstruation, or sexual skin size alone, provided little predictive power. In their laboratory study of regularly cycling female baboons, the maximum swelling size varied greatly between females, but was relatively constant within the cycles of any one female. This was also the case in the present sample of baboon females [HAUSFATER, 1974], at least for those females who showed regular sexual cycling. Thus, for each female there is a specific swelling size which may be considered her maximum, and achievement of which provides a prediction of proximity to ovulation.

Behavioral Indications of Proximity to Ovulation

At the onset of the study it was intended to obtain samples of each cycling female's behavior for an 11-day period surrounding ovulation. The samples were to start on cycle day D-7, and to continue through the third day following rapid deturgescence, cycle day D + 3. This procedure was adopted so as to bracket the presumed day of ovulation, cycle day D-2 or D-3, by several days in both directions as well as to allow statistical analysis of any behavioral changes that might occur with the onset of rapid deturgescence (see below). Originally, samples were begun on a female when she reached the swelling size that past records showed to be her maximum. If no past records were available on a female, then samples were routinely begun when a female showed a sexual skin swelling of 6 on the scale of 1–20. However, this procedure was not completely satisfactory. While, as noted above, cycle length and maximum swelling were relatively constant for females who cycled regularly, this was not the case for females who had recently resumed cycling, for example after the death of an infant or at the termination of lactation amenorrhea. Cycle length and maximum swelling size varied considerably for the first two or three cycles of these females.

However, two behavioral events proved to be good predictors of proximity to deturgescence both in females who cycled regularly and in those who did not do so. These two behavioral events were (1) the first observed occurrence of residual coagulated ejaculate on the female's perineum, an indication of recent mating, and (2) the first formation of a male-female consort relationship. The first occurrence of residual ejaculate on the female's perineum usually occurred eight days prior to rapid deturgescence, i.e. cycle day D-8, while the first consort relationship of the female was usually formed six days prior to deturgescence, i.e. cycle day D-6. Thus, after the first few months of the study, samples were begun on each female on the first day that she showed residual ejaculate on her perineum, and in general samples that bracketed the period of fertility were obtained consistently. However, sampling considerations aside, insofar as deturgescence is tied to ovulation, these events may be considered behavioral indications of the onset of estrus and of proximity to ovulation.

At least once per day, each female's perineum was visually inspected using 7 × binoculars at about 300 m distance and the presence of coagulated ejaculate was noted whenever observed.

Secretions of the prostate gland cause the semen to coagulate rapidly after ejaculation [VAN WAGENEN, 1936], and thus coagulated semen provides a relatively durative indication of recent mating activity. Only cycling females were seen to have coagulated ejaculate on their perinea. On the average, residual coagulated ejaculate first appeared on a cycling female's perineum on cycle day D-8 or D-7, and coagulated ejaculate was rarely observed on a cycling female's perineum after cycle day D + 2 (table XIX). The mean interval between first and last occurrence of coagulated ejaculate on the perinea of cycling females was 9.3 days, though within any one cycle, ejaculate seldom was present on the female's perineum on all days between the first and last occurrence (table XIX).

At the termination of each focal female sample, the presence and identity of any male consort of the female was scored. The identity and rank of the

Table XIX. First and last occurrence of residual coagulated ejaculate on perineum by female and cycle day. Ejaculate interval is defined as total number of days between first and last occurrence of ejaculate on perineum within a single cycle

Female	First occurrence of ejaculate			Last occurrence of ejaculate			Ejaculation interval (A + B)
	mean number cycle days pre-deturgescence (A)	SD	number of cycles	mean number of cycle days post-deturgescence (B)	SD	number of cycles	
Lulu	7.3	2.45	11	1.2	0.55	12	8.5
Judy	5.8	1.70	11	1.2	0.90	12	7.0
Oval	8.5	1.50	4	1.7	0.83	4	10.2
Alto	11.3	2.05	3	2.0	0.71	4	13.3
Fluff	9.7	5.76	4	5.7	6.60	3	15.4
Ring	9.3	4.50	3	1.3	1.25	3	10.6
Total, all females	7.7	3.42	36	1.7	2.34	38	9.4
All females except Lulu	7.9	3.75	25	1.9	2.77	26	9.8

male consorts, and the duration of individual consort relationships, will be discussed in detail in chapter IV. However, on the average, cycling females first formed consort relationships on cycle day D-6 and only rarely did a female continue to consort past the day of deturgescence (D-day). The mean interval between first and last consortship was 6.8 days (table XX), though as will be shown in chapter IV, in any one cycle a female usually was not in consort on all days between her first and last consort relationship.

In summary, although a definite external or behavioral indication of ovulation has yet to be found, the onset of estrus, i.e. cycle days D-7 through D-1 can be determined from achievement of maximum perineal swelling, consort formation, and presence of coagulated ejaculate on the female's perineum. Obviously, the onset of estrus also can be determined retrospectively once the onset of deturgescence is observed. Behavioral data to be presented below indicate that several marked changes in the rate of social interaction of cycling females did occur on cycle days D-2 and D-3, but the exact relationship of these behavioral and perineal changes to ovulation can be determined only in a laparoscope-equipped laboratory or with free-rang-

Table XX. First and last occurrence of consort relationship by female and cycle day. Consortship interval is defined as total number of days between first and last consortship within a single cycle

Female	First occurrence of consortship			Last occurrence of consortship			
	mean number of cycle days pre-deturgescence (A)	SD	number of cycles	mean number of cycle days post-deturgescence (B)	SD	number of cycles	Consortship interval (A+B)
Lulu	5.8	1.72	9	0.7	0.75	12	6.5
Judy	3.9	1.64	8	0.7	0.65	12	4.6
Oval	7.8	0.96	4	0.7	0.50	4	8.5
Alto	8.0	3.46	4	0.5	0.58	4	8.5
Fluff	7.2	4.57	4	1.0	0.82	4	8.2
Ring	5.7	4.04	3	1.3	0.58	3	7.0
Scar	8.0	0.00	1	0.0	0.00	1	8.0
Total, all females	6.1	2.81	33	0.7	0.67	40	6.8
All females except Lulu	6.2	3.14	24	0.7	0.65	28	6.9

ing animals telemetered for ovarian temperature or other physiological correlates of ovulation.

Estrous Behavior of Baboons

Review

BOPP [1953], ROWELL [1967, 1968] and SAAYMAN [1970, 1971a] have described cyclical changes in behavior accompanying the menstrual cycle of baboons. In these studies females usually have been classified as pregnant, suckling, cycling, or noncycling, and then cycling females further subdivided into those whose sexual skin was flat, inflating, swollen or deflating. Typically, a female is considered in the 'swollen' phase during the 7–10 days preceding deturgescence or whenever maximum turgescence is first achieved. On the basis of these studies, the results pertinent to 'swollen' females may be summarized as follows:

Swollen females received proportionally more mount attempts than did females in other cycle states and a larger proportion of mounts to swollen females were completed (i.e. copulations) than were mounts to females in other cycle states. Ejaculation by adult males occurred exclusively with swollen females [SAAYMAN, 1970].

Swollen females presented proportionally more to adult males than to younger males and presentation by swollen females brought proportionally more mount attempts than did presentations by females in other cycle states [SAAYMAN, 1970; ROWELL, 1967].

Adult males and swollen females groomed each other more frequently than would be expected if grooming were distributed randomly. In fact, adult males rarely groomed any other class of animals besides swollen females. Swollen females groomed adult males only slightly more than other classes of animals [SAAYMAN, 1970; ROWELL, 1968].

ROWELL [1967] reported that swollen anubis baboon *(Papio anubis)* females were involved in more aggressive interactions than females in other cycle states, while SAAYMAN [1972] concluded that chacma baboon *(P. ursinus)* females in the late deturgescent phase of the menstrual cycle were involved in significantly more aggressive interactions than were females in other phases of the menstrual cycle, including swollen females.

Analysis of Focal Female Samples

The focal animal sampling technique and behaviors recorded therein were discussed in detail in chapter I. The present section summarizes the behavior of estrous females as revealed by that sampling technique. Table XXI gives the number of minutes of focal sample time for each estrous female on cycle days D-7 through D+3. As noted above, this 11-day period of the baboon menstrual cycle brackets both the day of rapid deturgescence of the swollen sexual skin and the presumed day of ovulation.

Rate Calculation

The rate information about each female's behavior was drawn only from her own focal samples. A female's rate of interaction for a particular cycle day, e.g. cycle day D-3, was calculated from all focal samples on her on that cycle day in several menstrual cycles.

Let N_{ijk} equal the number of social interactions in which the ith female participated during focal samples on her on the jth cycle day of her kth sampled menstrual cycle. Furthermore, let t_{ijk} equal the total number of minutes of focal sample time in such samples. Then, L_{ij}, the ith female's mean rate of interaction on the jth cycle day, in all k of her sampled menstrual cycles combined, was calculated by the formula:

$$L_{ij} = \frac{\sum_{k} n_{ijk}}{\sum_{k} t_{ijk}} \tag{1}$$

i.e. the mean rate of interaction for the ith female on the jth cycle day of all sampled menstrual cycles was calculated as the total number of interactions that occurred during focal samples of her on that cycle day or all sampled menstrual cycles divided by the total number of sample minutes on her on that cycle day of all sampled menstrual cycles. Several assumptions about behavior were made for the purpose of this calculation and will be discussed below.

Once L_{ij} was calculated for each female for each cycle day from D-7 through D+3, a mean rate of interaction for all females for each cycle day was calculated by two different methods. In the first method, a mean rate (henceforth mean female rate), $\overline{L}_j$, of interaction for all females was calculated for each cycle day by the formula:

$$\overline{L}_j = \frac{\sum\limits_{i=1}^{n} L_{ij}}{n} \tag{2}$$

where n equals the total number of females who were sampled for any length of time on the jth cycle day in one or more menstrual cycles and L_{ij} is as above.

This mean female rate, $\overline{L}_j$, of interaction for a specific cycle day is influenced equally by each female's behavior regardless of the number of cycles or number of minutes for which she was sampled; in essence, it is the mean over all females of each female's rate. For example, females Lulu and Ring each potentially influenced this mean rate equally, though Lulu was under observation for substantially more cycles and more minutes than was Ring (table XXI). Similarly, the data on females Judy and Lulu, who both cycled throughout the study period did not 'swamp' the mean female rate for all females.

In addition to the mean female rate, a pooled mean rate, $\overline{L}'_j$, of interaction for all females combined was calculated for each cycle day from D-7 through D+3. The pooled mean rate was calculated for each cycle day by extending formula (1) above to the following:

$$L'_j = \frac{\sum\limits_{i,k} n_{ijk}}{\sum\limits_{i,k} t_{ijk}}, \tag{3}$$

i.e. the pooled mean rate, $\overline{L}'_j$, of interaction for all females on the jth cycle day of all sampled cycles was calculated as the total number of interactions in which all focal females participated on that cycle day of all sampled menstrual cycles divided by the total number of minutes in such samples.

Table XXI. Focal-female sample times, giving the number of minutes of focal sample time for each individual female by cycle day

Female	Cycle day										
	D-7	D-6	D-5	D-4	D-3	D-2	D-1	D	D+1	D+2	D+3
Lulu	360.00	572.11	590.03	639.82	748.30	875.25	866.89	808.58	568.67	629.55	497.05
Judy	289.88	345.57	485.14	506.93	544.77	623.45	745.76	631.43	528.16	553.69	378.93
Oval	240.00	259.43	224.09	318.38	369.12	300.00	288.69	287.37	156.58	180.00	120.00
Alto	239.12	180.00	295.88	213.14	278.46	276.10	213.00	230.47	226.04	259.00	175.26
Fluff	170.00	194.63	173.61	231.48	210.00	299.47	249.75	249.88	213.27	133.77	137.81
Ring	70.00	140.00	140.00	137.53	180.90	164.70	205.79	128.98	146.20	60.00	134.93
Total	1,369.00	1,691.74	1,908.75	2,047.28	2,331.55	2,538.97	2,569.88	2,336.71	1,838.92	1,816.10	1,443.98

In calculating the $\overline{L}'_j$s, it has been assumed that the observed cycles are essentially homogeneous with respect to rates of behavior, i. e. that they are independent samples from a single underlying Poisson process and that the underlying rate has remained constant through time [COX and LEWIS, 1966]. Additionally, it has been assumed that the rate of social interaction by adult female baboons depends primarily on their cycle day, rather than on factors particular to specific females or specific menstrual cycles. In calculating the $\overline{L}_j$s, it has been assumed that the cycles of any one female are essentially homogeneous, but that the various females may not be so. In fact, the actual numerical results produced by these two different methods of calculating a mean rate of social interaction for adult females differ only slightly.

The rates (per hour) at which individual females participated in twelve classes of social interactions (discussed below), as well as the raw data from which the mean rates were calculated, are given in HAUSFATER [1974], while the mean female rate of social interaction for these twelve classes of behavior, as well as the standard error, and standard deviation of the mean rate are summarized in table XXII. Additionally, figures 14–18 graphically present the pooled mean rate of twelve types of interaction (to be discussed below) for all females by cycle day and the 95% confidence intervals for the pooled mean rates [COX and LEWIS, 1966]. The significance of day-to-day changes in the pooled mean rates of social interaction for all females has been evaluated using the variance ratio [COX and LEWIS, 1966] at the 0.05 level of significance and the significant changes are marked by an asterisk in the figures. Of course, changes extending over several days may reach significant levels, even when the day-to-day changes do not.

Rates of Social Interaction. Cycle Days D-7 through D+3

As noted above, for each cycle day from D-7 through D+3 a pooled mean rate of participation for all females in each of twelve classes of social interaction was calculated and the day-to-day changes in these pooled mean rates will be discussed in the remainder of this chapter. Thus, in contrast to previous studies of estrous behavior, reviewed above, that have analyzed changes in female behavior over the entire menstrual cycle, the present analysis gives detailed treatment to changes in female behavior over an extremely restricted, but biologically significant, portion of the menstrual cycle, the period surrounding ovulation.

The individual and mean rates for 3 classes of social interaction, namely (1) agonistic bouts won by the estrous female, (2) agonistic bouts in which the estrous female was defeated, exclusive of herding by mature males, and (3)

Table XXII. Mean female rates of participation by estrous females in various categories of social interaction [HAUSFATER, 1974] as well as standard error (SE) of the mean rate and standard deviation (SD) of the sample distribution. See text for method of calculation

Category of social interaction	D-7			D-6			D-5			D-4			D-3		
	mean	SE	SD	mean	SE	SD	mean	SE	SD	mean	SE	SD	mean	SE	SD
Agonistic behaviors															
1 Agonistic bouts won by female	0.55	0.31	0.77	0.46	0.16	0.39	0.42	0.15	0.37	0.79	0.37	0.92	0.35	0.21	0.52
2 Agonistic bouts lost by female	1.80	0.57	1.40	1.10	0.39	0.96	0.92	0.34	0.83	0.88	0.25	0.61	1.01	0.22	0.55
3 Herding	0.42	0.23	0.57	0.98	0.43	1.04	1.20	0.59	1.44	1.69	0.62	1.52	1.43	0.36	0.89
4 Undecided agonistic bouts	0.00	0.00	0.00	0.00	0.00	0.00	0.00	0.00	0.00	0.00	0.00	0.00	0.01	0.01	0.03
5 All agonistic interactions	2.76	0.63	1.53	2.54	0.49	1.21	2.54	0.63	1.55	3.36	0.79	1.94	2.81	0.35	0.86
Nonagonistic Behaviors															
1 Mounting by mature males	1.17	0.20	0.50	1.36	0.29	0.70	1.98	0.61	1.49	1.00	0.28	0.68	1.45	0.37	0.90
2 Incipient mounts	0.43	0.17	0.43	0.23	0.13	0.31	0.30	0.25	0.61	0.82	0.50	1.24	0.70	0.37	0.89
3 Mounts by immature males	0.12	0.08	0.19	0.48	0.18	0.43	0.21	0.12	0.29	0.72	0.48	1.16	0.13	0.06	0.15
4 Presentation	0.97	0.41	1.00	0.91	0.15	0.35	0.37	0.14	0.35	0.49	0.10	0.25	0.91	0.40	0.97
5 Perineal inspection	1.90	0.28	0.69	1.69	0.22	0.55	1.44	0.40	0.97	1.33	0.36	0.89	1.62	0.22	0.54
6 Pelvis grasping	0.47	0.26	0.64	0.63	0.28	0.68	0.58	0.35	0.85	0.99	0.22	0.54	1.44	0.26	0.64
7 Grooming	0.84	0.21	0.51	1.78	0.51	1.25	2.62	1.11	2.71	2.13	0.50	1.23	1.43	0.38	0.93
8 Following	4.60	0.74	1.81	7.90	1.30	3.19	8.26	1.45	3.56	9.38	1.55	3.79	11.24	1.19	2.91
9 All nonagonistic interaction	10.52	1.29	3.16	15.02	2.12	5.19	15.76	2.66	6.52	16.87	2.00	4.91	18.91	1.35	3.32
10 All social interaction	13.28	1.28	3.14	17.56	2.25	5.50	18.30	3.20	7.85	20.23	2.62	6.42	21.72	1.51	3.70

Table XXII (continued)

Category of social interaction	D-2 mean	SE	SD	D-1 mean	SE	SD	D mean	SE	SD	D+1 mean	SE	SD	D+2 mean	SE	SD	D+3 mean	SE	SD
Agonistic behaviors																		
1	0.42	0.15	0.37	0.33	0.11	0.28	0.50	0.14	0.35	0.49	0.20	0.50	0.89	0.49	1.20	0.89	0.40	0.97
2	0.64	0.12	0.30	0.95	0.38	0.92	1.09	0.24	0.59	1.27	0.37	0.92	1.87	0.35	0.86	2.23	0.47	1.16
3	1.89	0.41	0.99	1.04	0.24	0.59	0.04	0.04	0.11	0.02	0.02	0.04	0.17	0.15	0.37	0.08	0.08	0.20
4	0.00	0.00	0.00	0.00	0.00	0.00	0.00	0.00	0.00	0.00	0.00	0.00	0.04	0.04	0.09	0.03	0.03	0.06
5	2.96	0.39	0.95	2.33	0.53	1.30	1.48	0.24	0.59	1.78	0.28	0.68	2.61	0.75	1.84	3.26	0.58	1.42
Nonagonistic behaviors																		
1	1.73	0.49	1.21	1.13	0.16	0.39	1.07	0.31	0.77	0.85	0.28	0.68	0.55	0.17	0.41	0.81	0.17	0.42
2	1.13	0.48	1.18	0.78	0.40	1.00	0.38	0.14	0.34	0.07	0.07	0.17	0.02	0.02	0.04	0.12	0.07	0.17
3	0.09	0.06	0.15	0.19	0.08	0.19	0.36	0.08	0.19	0.52	0.19	0.45	0.15	0.11	0.26	0.22	0.14	0.35
4	0.48	0.08	0.21	0.32	0.13	0.32	0.70	0.16	0.39	1.39	0.54	1.31	0.86	0.19	0.46	0.52	0.16	0.39
5	1.69	0.28	0.67	1.52	0.32	0.78	1.77	0.33	0.81	2.07	0.51	1.24	1.13	0.36	0.88	0.70	0.14	0.33
6	1.52	0.37	0.91	1.21	0.36	0.88	0.73	0.26	0.63	0.97	0.24	0.58	0.34	0.15	0.36	0.13	0.07	0.16
7	2.26	0.58	1.42	2.31	0.36	0.87	2.11	0.53	1.31	0.88	0.19	0.46	0.48	0.33	0.81	0.66	0.50	1.22
8	10.90	1.63	4.00	9.04	1.02	2.51	5.63	0.75	1.84	1.96	0.50	1.24	0.69	0.34	0.82	0.46	0.14	0.35
9	19.81	2.42	5.93	16.50	1.48	3.62	12.75	1.03	2.53	8.70	1.76	4.31	4.22	0.71	1.74	3.62	0.93	2.27
10	22.76	2.78	6.80	18.84	1.96	4.80	14.38	1.20	2.95	10.47	1.71	4.19	7.19	1.17	2.86	6.85	0.17	1.73

undecided agonistic bouts, were based on bouts between the female and any other individual in the group (fig. 15, table XXII). In another eight classes of social interaction (see below), analysis was restricted to interactions between the female and males of the J2 class or older. Males in these age classes produce viable sperm [HENDRICKX and KRAEMER, 1969], and in what follows, will be referred to as 'mature males'.

Interactions for which only participation between estrous females and mature males (males of the J2 class or older) was considered included: (4) male grasped pelvis of female, (5) male incipient mounted female, (6) male mounted female, with or without ejaculation, (7) female presented hindquarters to male, (8) male inspected female's perineum, (9) male followed female, (10) female groomed, or was groomed by, male, and (11) male herded female (fig. 15–18, table XXII). Finally, analysis of one remaining category of social interaction (12) immature male mounted estrous female (fig. 16, table XXII), was restricted to participation by males of the J1 class or younger, i.e. sexually immature males (fig. 6).

All Social Interactions

Figure 14 shows the pooled mean rate of participation by estrous females in all categories of social interaction combined ('all social interaction') on

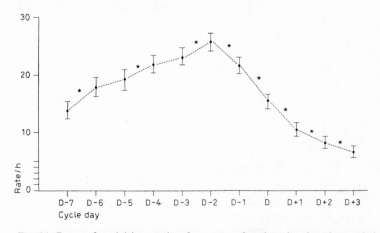

Fig. 14. Rates of social interaction for estrous females, showing the pooled mean rate/h of all social activity for estrous females as calculated from focal female samples by a method described in the text. Vertical lines indicate the 95% confidence intervals for the mean rate and a star indicates day-to-day changes in the mean rate that were significant at the 0.05 level. Method of calculating confidence intervals and significance testing are discussed in the text.

cycle days D-7 through D + 3 and the 95% confidence intervals for the pooled mean rates. As shown there, the pooled mean rate of all social interaction for cycling females increased steadily from D-7 to a peak at D-2, then declined rapidly through cycle days D + 2 and D + 3. Presumably, the mean rate of all social interaction shown for cycle day D + 3 is close to the minimum or baseline level for cycling females. For the eleven cycle days shown, the day-to-day changes in the pooled mean rate of all social interaction were significant for all pairs of days but D-6 to D-5 and D-4 to D-3.

Agonistic Behavior of Estrous Females

Figure 15 presents the pooled mean rate at which estrous females won agonistic bouts from other animals of all ages and sexes on cycle days D-7 through D + 3. The pooled mean rate at which females won agonistic bouts was relatively uniform on the days preceding deturgescence, but increased significantly on the first two days after the onset of rapid deturgescence. The pooled mean rate at which estrous females were defeated in agonistic bouts, exclusive of herding by mature males, is also shown in figure 15. The higher rates of defeat on cycle days D-7 and D + 2 and D + 3 when compared to the intermediate cycle days indicate that this rate was depressed throughout the period of fertile matings. In fact, the rate at which estrous females were defeated in agonistic bouts reached a local minimum on cycle day D-2, close to the presumed time of ovulation. Table XXII summarizes the available information on undecided agonistic bouts between estrous females and individuals of all classes.

In sum, the above data indicate that the rate at which estrous females defeated, or were defeated by, other individuals in agonistic bouts, exclusive of herding by mature males, was lower on the three or four cycle days surrounding ovulation than on either the preceding or subsequent cycle days. These results are explained primarily by the fact that herding (see below) of the female by adult male consorts was the most frequent category of agonistic interaction by estrous females and had the effect of reducing the probability of interaction, both agonistic and nonagonistic, between the female and all other individuals in the group.

Herding

Figure 15 also shows the pooled rate of herding of females by mature males on the cycle days under consideration. Herding of females by males is considered particularly important in the formation of one-male units of hamadryas baboons *(Papio hamadryas)* [KUMMER, 1968]. However, in con-

trast to hamadryas baboons, yellow baboon *(P. cynocephalus)* males herd
females by nipping them on the thigh, flank, or rump, and only rarely on the
back of the neck as is typical of hamadryas. Male consorts among yellow
baboons herded females at all times of the day, but were especially active at
herding when other males were nearby. The usual result of herding was to
drive the female to the periphery of the group and away from all males but
the consort. This spatial peripherality of consort pairs was noted by DEVORE
[1965]. Thus, male consorts, regardless of rank, herded females away
from other males, rather than chasing competing males away from the
female.

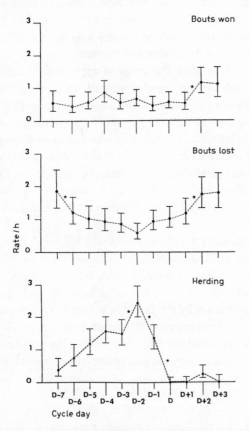

Fig. 15. Rates of agonistic bouts for estrous females, showing the pooled mean
rates/h of agonistic bouts won by focal female, agonistic bouts lost by focal female, exclusive of herding by mature males, and herding by mature males as calculated from
focal female samples by a method described in the text. Symbols as in figure 14.

In the period from D-7 through D + 3, the pooled mean rate of herding of estrous females by mature males increased from a low rate on cycle day D-7 to a peak four times as great on cycle day D-2; and the increase in the rate from cycle day D-3 to D-2 was statistically significant. With the onset of rapid deturgescence, the rate of herding of estrous females underwent a series of significant decreases and rapidly approached zero.

Nonagonistic Interactions. Mounting

Figure 16 shows the rate of mounting (fig. 4) and incipient mounting (fig. 5) of estrous females by sexually mature males. The pooled mean rate

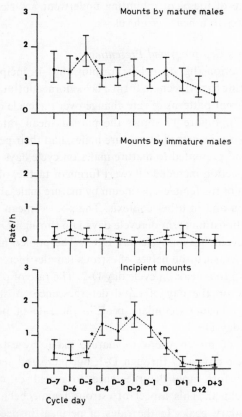

Fig. 16. Rates of mounting and related behavior for estrous females, showing the pooled mean rates/h of mounts by mature males, mounts by immature males on estrous females, and incipient mounts by mature males as calculated from the focal female samples by a method described in the text. Symbols as in figure 14.

of mounting, both with and without ejaculation, was highest on cycle day D-5 and was uniform at a rate of just over 1 mounting/h on all other cycle days prior to the onset of deturgescence. After D-day, the rate of mounting of females by mature males declined slowly with no significant day-to-day changes. The pooled rate of mounting of females by immature males (fig. 6, 16) was low on all cycle days prior to deturgescence, approaching zero on cycle day D-2. Thus, sexually mature males and sexually immature males mounted cycling females at very different times in the menstrual cycle. The pooled mean rate of incipient mounting of females by mature males (fig. 16) showed a significant increase between cycle days D-3 and D-2 with the peak rate of incipient mounting occurring on cycle day D-2. After the onset of rapid deturgescence, the rate of incipient mounting underwent a series of significant and rapid decreases to a near zero level.

Perineal Inspection, Pelvis Grasping, and Presenting

Although perineal inspection (fig. 4), pelvis grasping (fig. 4), incipient mounting (fig. 5), and mounting (fig. 4) seem to form a behavioral continuum, these behaviors showed different patterns of rate change over the cycle days being considered (fig. 16, 17). Figure 17 shows the pooled mean rate at which females had their perinea inspected by mature males, had their pelves grasped by mature males, and presented to mature males on cycle days D-7 through D + 3. Perineal inspection included all overt forms of tactile, olfactory, and visual examination of the female's perineum by mature males, both in response to a presentation and in other contexts. The pooled mean rate of perineal inspection was almost uniform on cycle days D-6 through D + 1, though with a slight peak occurring on cycle day D + 1. The pooled mean rate at which mature males grasped the pelves of estrous females increased smoothly from cycle day D-5 to a peak on cycle day D-2. The rate of pelvis grasping declined markedly on the day of rapid deturgescence. A slight (though not statistically significant) secondary peak in the rate of pelvis grasping occurred on cycle day D + 1.

The pooled mean rate of presentations to mature males by estrous females was uniform on cycle days D-7 through D-1, but increased significantly, beginning the day before rapid deturgescence, to a peak on cycle day D + 1, then rapidly declined. This aspect of estrous female behavior explains, in part, the secondary peaks in the rates of perineal inspection (fig. 17), pelvis grasping, (fig. 17), and mounts by immatures (fig. 16) seen on cycle day D + 1 also. On cycle day D + 1, females moved freely about the group, unhindered by male consorts. Females presented frequently to males

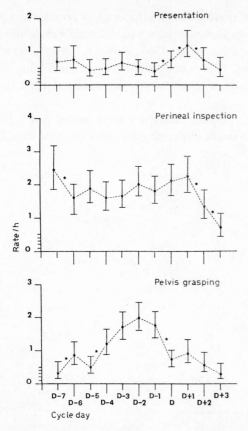

Fig. 17. Rates of presentation, perineal inspection, and pelvis grasping for estrous females; showing the pooled mean rate/h of presenting, perineal inspection, and pelvis grasping for estrous females as calculated from focal female samples by a method described in the text. Symbols as in figure 14.

of all ages on this cycle day (fig. 17), and the response of mature males to a presentation by a newly deturgescent female was a perineal inspection or a pelvis grasp (fig. 17), but seldom a mount or even an incipient mount (fig. 16). In contrast, immature males responded to newly deturgescent females with mounting behavior (fig. 16).

Grooming and Following

The two remaining classes of nonagonistic interaction to be discussed include following (fig. 5) of estrous females by mature males and grooming between estrous females and mature males. Grooming of females by adult males and vice versa has been shown to vary with a female's cycle state in several primate species [MICHAEL *et al.*, 1966; ROWELL, 1968; CHALMERS and ROWELL, 1971; SAAYMAN, 1971a] and male-female grooming has been used as an indication of estrus in rhesus monkeys, *Macaca mulatta* [LOY, 1971].

In the present study, grooming between mature males and estrous females (fig. 18) showed significant day-to-day increases on cycle days D-7

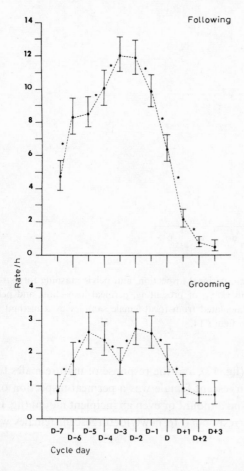

Fig. 18. Rates of following and grooming of estrous females; showing the pooled mean rates/h of following and grooming for estrous females as calculated from focal female samples by a method described in the text. Symbols as in figure 14.

through D-5, but decreased significantly on cycle day D-3, the presumed day of ovulation. The pooled mean rate of grooming reached a peak on cycle day D-2 then declined significantly again with the onset of deturgescence. The findings of this study agree with those of SAAYMAN [1971 a] that adult males only rarely groomed anestrous females or any other class of individuals.

The pooled mean rate of following of estrous females by mature males increased steadily from cycle day D-7 to a peak on cycle day D-3. With the onset of rapid deturgescence, the rate of following underwent a series of significant day-to-day decreases and approached a near zero rate by cycle day D + 3. The fact that mature males continued to follow estrous females at a high rate on cycle day D-3 suggests that female attractiveness and male preferences had not been particularly altered on this cycle day. Thus, the significant decline in the rate of grooming between mature males and estrous females on cycle day D-3 may be the result of some change in the female's *receptiveness* to the male, rather than a change in her *attractiveness* to the male or a change in adult male *preferences*. Further analysis of these data to separate changes in female attractiveness and receptivity from changes in male preferences during the above cycle days is anticipated.

Summary

This chapter provides necessary background information on the length of the menstrual cycle in baboons, the timing of ovulation within the cycle, the optimal time for mating, and the behavior of estrous females. This information will be used subsequently in a test of the priority-of-access model of mating behavior and in a discussion of mating patterns in Amboseli baboons. The following is a summary of the most important information in this chapter:

1. The length of the menstrual cycle of females in the study group, determined from the interval between successive onsets of deturgescence of the swollen sexual skin in 34 pairs of successive cycles, was 32.5 days. These data are from a sample of six females, though a total of eleven females underwent some portion of a menstrual cycle during the study period.

2. The hormonal basis of sexual skin changes was reviewed. The turgescent phase of the sexual skin cycle, from the first day of menstruation up to the day of onset of rapid deturgescence (D-day), corresponds to the phase of follicular activity in the ovarian cycle. The deturgescent phase of the sexual

skin cycle from D-day up to the first day of the subsequent menstruation, corresponds to the phase of luteal activity in the ovarian cycle. The mean length of the turgescent phase of the menstrual cycles of females observed during this study was 18.5 days (N = 35 cycles), while the mean length of the deturgescent phase of the menstrual cycle was found to be 13.9 days (N = 31 cycles).

3. In the laboratory, it has been shown, that fertile matings in baboons can occur from cycle day D-7 through D-1. However, cycle day D-3 is the optimal time for mating, and is probably the day of ovulation. It is also probable that under natural conditions, most fertile matings occur on, or very close to, cycle day D-3. Behavioral and perineal indications of proximity to ovulation were discussed. The first appearance of residual coagulated ejaculate on the perineum of a cycling female, and the first formation of a male-female consort relationship, proved to be particularly useful indications of the onset of the period of fertile matings.

4. The rates of social interaction of estrous females were discussed. The rate of all social interactions combined for estrous females increased steadily from cycle day D-7 through D-2, then declined rapidly with the onset of rapid deturgescence.

5. The rate of agonistic bouts between estrous females and individuals other than their male consorts was depressed throughout the period of fertile matings. The rate of herding of estrous females by mature males showed the same pattern of rate changes seen for all of the females' social activities.

6. The mean rate of mounting of estrous females by mature males was uniform throughout the fertile period while incipient mounting showed a peak on cycle day D-2. The mean rate of mounting of females by immature males was uniformly low during the period of maximum fertility, demonstrating that males of different ages had access to females at different times in the menstrual cycle.

7. Females presented to mature males most frequently on cycle day D+1 and the response of a mature male at that time frequently was a perineal inspection of the female or a pelvis grasp to the female, but only rarely a mount or incipient mount. In contrast, immature males responded to presentation from a female on cycle day D+1 with mounting.

8. Mature males and estrous females groomed each other at a high rate throughout the period of sexual activity, though a sharp decline in the rate of grooming occurred on cycle day D-3, the presumed day of ovulation. The rate at which mature males followed estrous females increased steadily

and did not show a decline on cycle day D-3. It is hypothesized that the receptivity of estrous females to mature males is lowered on cycle day D-3, the presumed day of ovulation, but that no such reduction occurs in the female's attractiveness to mature males or in the male's preference for the female. This hypothesis is open to verification with both field and laboratory data.

IV. The Mating System of Amboseli Baboons

Introduction

What determines the mating patterns in a baboon group? Do baboons mate preferentially with respect to dominance rank? Will the most dominant individual in a baboon group leave more offspring than any of the subordinates? The answers to these and related questions have strong implications for the genetic history of a baboon social group. Differential reproduction and its evolutionary importance was reviewed in chapter I.

In this chapter, the mating patterns in the study group of Amboseli baboons will be described and the relationship between dominance rank in males and access to estrous females analyzed. As one portion of this discussion, a priority-of-access model of mating behavior will be evaluated in the light of data from Alto's group.

The Formal Model

ALTMANN [1962] pointed out that if the proportion of the menstrual cycle during which macaque females are in estrus is constant among all females, and if females cycle independently of each other, then the probability, P_r, that at least r females will be in estrus simultaneously can be calculated from the cumulative binominal probability distribution:

$$P_r = \sum_{x=r}^{n} \binom{n}{x} p^x (1-p)^{n-x},$$

where r equals the number of females in estrus simultaneously, n equals the total number of cycling females in the group, and p equals the probability that a cycling female will be in estrus as opposed to some other cycle state.

Furthermore, if it is assumed that an estrous female will form a consort relationship, i.e. a temporary pair-bond, with the higher ranking of any two potential male partners, and that the consort relationship excludes all other potential mates for both partners, then a male of dominance rank r would be expected to be in consort with an estrous female only when at least r females are simultaneously in estrus in the group. For example, the first ranking male of a baboon group would be expected to consort whenever at least one estrous female was present. The second and third ranking males would be expected to consort with estrous females only when at least two and at least three estrous females were present in the group, respectively. Therefore, P_r above not only gives the probability that at least r females will be in estrus simultaneously, but also the probability that a male of dominance rank r will consort with an estrous female, versus the probability that he will not be in consort with an estrous female.

The above assumptions and their logical conclusions constitute the priority-of-access model of mating behavior as originally published by ALTMANN [1962] and the reader is referred to that work for a more detailed presentation. However, it should be obvious that since the p of the model is the proportion of time that females spend in estrus, and thus are assumed by the model to be in consort, then P_r can be taken to be the proportion of time that the *rth*-ranking male will be in consort with an estrous female. Thus, this priority-of-access model makes predictions about the amount of time that males of various ranks will have exclusive access to, i.e. will consort with, an estrous female.

Previous Tests of the Model

ALTMANN [1962] made it clear that he did not have sufficient data to test this model, though later SUAREZ and ACKERMAN [1971] were able to test the model using data provided by CARPENTER [1942a, b], KAUFMANN [1965] and CONAWAY and KOFORD [1965] from the Cayo Santiago rhesus monkey colony in Puerto Rico.

SUAREZ and ACKERMAN [1971] used as measures of access or reproductive success of males: (1) total days of copulation, (2) total days of consortship, and (3) number of different females mated. Since the model as originally published makes predictions about the amount of time that males of each rank will engage in consortship or other mating activity, the first two measures of mating success provide adequate tests of the model. The third measure of mating success, number of different females mated, does not provide a legitimate test of the model unless additional assumptions are made about the rotation of females among ranks during estrus.

Furthermore, the exact definition of these measures of mating success was not explicitly stated nor was the definition of estrus. This could be critical since LOY [1970] has shown that female macaques may exhibit 'estrous' behavior around menstruation when they are presumably non-fertile. Also the lack of systematic sampling of behavior in these earlier studies means that one cannot assume that the observed data constitute a random sample of male-female consort pairs. ALTMANN [1962], SADE [1968], and LOY [1971] have all noted that on Cayo Santiago copulations between lower ranking males and estrous females may take place in concealment, thereby minimizing interruption by higher ranking males. Nevertheless, the priority-of-access model provided an adequate fit to the observed data in five out of ten tests completed by SUAREZ and ACKERMAN.

Reproductive Success and its Measurement

The present chapter has two equally important goals: (1) to describe the mating system of Amboseli baboons, particularly the relationship between dominance and reproductive success in males, and (2) to test the priority-of-access model of mating behavior. The priority-of-access model as originally published makes predictions only about the amount of time that males of each dominance rank will be in consort with estrous females. Thus, if the two goals of the present study are both to be fulfilled by a test of the model alone, it must be shown that time spent in consortship is a good measure of reproductive success in baboons or an alternative measure must be chosen and the model revised accordingly.

Techniques for directly determining the contribution of specific males to subsequent generations, i.e. their reproductive success or fitness, are available. Primarily, these are techniques based on paternity determination of infants by genetic analysis, for example, chromosomal matching or analysis of blood serum proteins [SADE, personal commun.]. In species in which a female mates exclusively with one male per estrus and in which the gestation period is known, back-dating from the date of an infant's birth to the estrus that resulted in conception may also provide a method of paternity determination [CONAWAY and KOFORD, 1965]. However, in a purely observational study such as the present work, and in a population in which females mate with more than one male per estrus, some more indirect measure of the reproductive success of males of each dominance rank relative to males of all other ranks had to be employed.

For several reasons (discussed below), time spent in consortship does not provide an adequate measure of reproductive success in Amboseli baboons, or even an adequate measure of mating success in the sense of frequency of copulation. The two primary reasons for rejecting time spent in consortship as a measure of reproductive success in Amboseli baboons were that (1) many matings between mature males and fertile females occurred outside of consortships, and (2) once a consortship was established, males actually mated with estrous females at a highly variable rate. Thus, a mature male may never have been in consort with an estrous female, but still have been successful in mating with her, or two males may have spent equal amounts of time in consort with an estrous female, but shown quite different frequencies of mating with her, i.e. different mating successes. The precise nature of the consort relationship is discussed in detail in subsequent sections of this chapter.

The measure of reproductive success that was eventually chosen for use in the present study was the normalized rate of mounting with ejaculation, i.e. copulation, by sexually mature males of each rank with females on or near the optimal cycle day for mating. Normalization over all ranks of the rates of copulation by mature males with estrous females yielded a measure of each rank's mating success relative to all other ranks and, presumably, of its reproductive success or fitness as well. In essence, the normalized rate gives the proportion of all copulations in which the male partner was of a specific rank, and thus an estimate of the probability that a male of a specific rank was the copulation partner of a female given that a copulation occurred. As noted in chapter I, a male-female mounting was considered a copulation only if the male exhibited an ejaculatory pause at the termination of the mounting or if the female had fresh ejaculate on her perineum immediately after the mounting. Also as indicated in the last chapter, males of the J2 age class, i.e. 2–4 years of age or older, were considered sexually mature [HENDRICKX and KRAEMER, 1969].

However, use of normalized rates of copulation as a measure of reproductive success necessitated a change in the original model, or strictly speaking, an additional assumption. Specifically, the assumption is made that males of the rth rank will carry out copulations in proportion to the time that at least r females are simultaneously in estrus. As noted above the model as originally published predicts access time or proportion of time spent in consort by males of each rank, rather than normalized rates of copulation. Thus, in the present case, time spent in the act of copulation might be a more appropriate measure of reproductive success for use in testing the model than normalized frequency of copulation per unit time. However, it is obvious that for brief events with only a small variance in duration, such as a copulation in baboons which typically requires less than 1 min, the normalized rates will be very nearly the same as the normalized proportion of time spent in copulation.

Calculation of Normalized Rates of Copulation

A copulation in baboons requires both a male and a female partner. Consequently, information on the rate of copulation between a specific female and males of rank r can be obtained both from focal samples on the female and from focal samples on the males of that rank, and both types of samples were used in the present analysis. When three males had triangular dominance relationships and thus were all assigned the same rank number, information about the rate of copulation between the female and males of rank r was obtained from samples on the female and from samples on each of the three males. This method differs from that used in chapter III in which only information in the focal female samples was used to estimate rates of social interaction for estrous females.

For each cycle day (D-7, D-6, etc.) of a specified female, let: n_{rj1} = the number of copulations between her and males of the r*th* rank during her focal samples on that day of her j*th* menstrual cycle; n_{rj2} = the comparable number of copulations with her that were observed in all focal samples taken on males of rank r on that day of her j*th* menstrual cycle; t_{rj1} = the total number of focal sample minutes obtained on the female on that day of her j*th* menstrual cycle whenever at least one sexually mature male occupied the r*th* rank; t_{rj2} = the comparable number of focal sample minutes obtained on males of rank r on that cycle day of her j*th* menstrual cycle; and m_{rj} = the number of males that jointly occupied the r*th* rank on that day of her j*th* menstrual cycle.

Then the rate, L_r, of copulation between this female and r*th* ranking males on that cycle day was calculated as:

$$L_r = \frac{\text{number of copulations}}{\text{focal sample minutes}} = \frac{\sum_j (n_{rj1} + n_{rj2})}{\sum_j (m_{rj} t_{rj1} + t_{rj2})}. \tag{4}$$

It is helpful to think of the present analysis in terms of male-female pairs, rather than in terms of male or female behavior alone. Although only one female per sample was the focal individual, if three males occupied rank r during a sample, then the female's sample provided information on three pairs, each including the female and one of the r*th* ranking males. Thus, it is obviously necessary to multiply the total number of sample minutes, t_{rj1}, on the female by the number of males, m_{rj}, who occupied the r*th* rank at that time. In most instances, m_{rj} equals one, or zero, i.e. for those cases in which a mature male did not occupy the r*th* rank, but in the case of the brief periods of triangular dominance relationship between males Max, Cowlick, and Sinister (fig. 9), $m_{rj} = 3$ for rank seven and $m_{rj} = 0$ for ranks six and eight. In fact, males of ranks six through eight actually had a zero expected probability of copulation with estrous females, but had this not been the case, an alternative model of mating behavior would have been required: the priority-of-access model implicitly assumes a linear dominance order among males.

A separate L_r was calculated for each individual female-male rank pair for cycle days D-7 through D-1 of the female's menstrual cycles. Then for each of the above cycle days, a pooled rate, $\overline{L}'_r$, of copulation between males of rank r and all cycling females was calculated by extending formula 4 so that summation was done both over all menstrual cycles and over all individual females as well; this procedure is analogous to the extension of formula 1 into formula 3 as was done in chapter III. The reader is referred to chapter III for an explanation of the formulae and a discussion of related assumptions.

Finally, the $\overline{L}_r$'s for all ranks on each cycle day were normalized:

$$C_r = \frac{\overline{L}'_r}{\sum_r \overline{L}'_r},$$

where C_r equals the observed proportion, or estimated probability, of a copulation between a male of rank r and an estrous female on a specified cycle day, given that a copulation occurred. When the C_r's for all ranks are added, they sum to one, and are therefore not independent of each other, e.g. as the value of C_r for any specific rank increases, the total of the C_r's for all other ranks decreases and by the same amount. Therefore, the proportion of all copulations, C_r, attributable to males of each rank effectively measures each rank's reproductive success in comparison to all other ranks.

Calculation of Expected Values
The mean duration of all menstrual cycles recorded during this study was 32.5 days (table XVIII, female Lulu's data included), and the duration of estrus, i.e. cycle day D-7 through D-1 inclusive (chapter III), was seven days. (Of course, as noted in chapter III, a mating on cycle day D-3 has a higher probability of resulting in conception than does a mating on any other cycle day). Thus, the probability that a cycling female was in estrus rather than some other cycle state was 7/32.5 or 0.21.

A female was defined as cycling from the first day of menstruation immediately preceding the onset of regular sexual skin swelling. However, if no menstruation was observed, the female was considered cycling as of the first day she exhibited a swelling above her rest level. A female was considered noncycling as of cycle day $D + 13$ of any cycle which was not followed by renewed swelling in at least 30 days, regardless of whether swelling ceased due to pregnancy or to other causes. No sexual skin swelling during pregnancy was exhibited by any of the females in the study group, though chacma baboons *(Papio ursinus)* are reported to show small swellings throughout gestation [GILLMAN and GILBERT, 1946; ALTMANN, 1970]. On every day of the study period, 2–6 cycling females were present in Alto's group.

The probability of obtaining at least one estrous female, at least two estrous females, etc., to a maximum of six estrous females on any day of study was read from a table of the cumulative binomial probability distribution [Harvard University Computation Laboratory, 1955] entered with p =

Table XXIII. Expected number of females in estrus, giving the cumulative probabilities of obtaining various numbers of females simultaneously in estrus when from one to six cycling females were present in the group, as calculated from the cumulative binomial probability distribution entered with $p = 0.21$, and comparable probabilities calculated from the actual number of study days that one or more females in the group were in estrus

Number of cycling females	Number of study days	Probability of r females simultaneously in estrus where r equals					
		1	2	3	4	5	6
A. Expected under cumulative binomial probability distribution, p = 0.21							
1	0	0.210	–	–	–	–	–
2	84	0.376	0.044	–	–	–	–
3	148	0.507	0.144	0.009	–	–	–
4	62	0.610	0.196	0.031	0.002	–	–
5	92	0.692	0.283	0.066	0.008	0.000	–
6	14	0.756	0.369	0.111	0.020	0.002	0.000
Mean, weighted by days		0.547	0.160	0.027	0.003	0.000	0.000
Normalized probabilities		0.742	0.217	0.037	0.004	0.000	0.000
B. Observed data							
		Number of females simultaneously in estrus					
Number of days (N = 400)		188	46	8	0	0	0
Cumulative probabilities		0.605	0.135	0.020	0.000	0.000	0.000
Normalized probabilities		0.796	0.178	0.026	0.000	0.000	0.000

0.21 (see above) and the number of 'trials', n, equal to the number of cycling females in the group on that day. Table XXIII gives these cumulative probabilities for days with 1–6 cycling females in the group as well as the number of such days in the study period.

For the entire study period, the cumulative probability, P'_r, of obtaining at least r estrous females was estimated in two ways. First these probabilities were estimated by the formula:

$$P'_r = \frac{\sum_n D_n \cdot P\,(x \geqslant r \mid n, p)}{400},$$

where $P\,(x \geqslant r \mid n, p)$ is the probability of obtaining at least r estrous females out of n cycling females that spend a proportion, p, of each cycle in estrus and D_n equals the number of days on which n cycling females were present in the group. The denominator, 400, is, of course, the duration of the study in days.

Table XXIII lists the cumulative probabilities of obtaining from one to six estrous females in the group on any study day as estimated by the above method. Also listed in table XXIII is a second set of estimated cumulative probabilities, calculated from the actual frequency of such days during

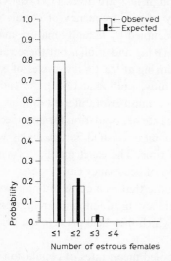

Fig. 19. Number of females in estrus; showing the observed and expected distribution of number of females simultaneously in estrus as calculated from the cumulative binomial probability distribution and from actual field data. Estrus is defined as cycle days D-7 through D-1 inclusive for this analysis. The two distributions do not differ significantly at the 0.05 level; see text for further explanation.

the study as determined from the field records. These two cumulative probability distributions were normalized and are presented graphically in figure 19. The two distributions shown in figure 19 do not differ significantly from each other ($\chi^2 = 1.82$; d.f. = 3; 0.05 level of significance) nor do the noncumulative distributions. This test supports indirectly the assumption that females cycle independently of each other. In the tests of the priority-of-access model which follow, the normalized cumulative probability distribution based on the actual field data were used as expected values for male reproductive success rather than the values calculated directly from the cumulative binomial probability distribution. This choice actually changed the expected probability of a completed copulation for any rank only slightly.

Results

Rank-Specific Rates of Copulation

The rates of copulation between the six individual females whose behavior was sampled in one or more menstrual cycle and sexually mature males of all ranks on cycle days D-7 through D-1 are given in HAUSFATER [1974] as are procedures for the recovery of raw frequency of copulation data. Dominance ranks 12–14 were not occupied by sexually mature males during the sampled cycles of two females (Ring and Fluff), but these ranks were occupied by sexually mature males during at least a few cycles of each of the four other cycling females (Alto, Judy, Lulu, and Oval). As shown in chapter II, most mature males occupied a number of different dominance ranks during the study period and thus the rate of copulation for any specific rank was the result of the behavior of a number of different males who occupied that rank for varying periods of time. The exact number of males who occupied each rank and their duration of occupancy can be determined from figures 9 and 10. It is important to note that as a convention, pairs of males undergoing a period of inconsistency in dominance relationship (table XIV) were each assigned the ranks which they held prior to the period of inconsistency.

Table XXIV lists the rank-specific pooled mean rates of copulation between sexually mature males of each rank and all cycling females combined on cycle days D-7 through D-1. Also given in table XXIV is the exact number of minutes that every possible pair – composed of a sexually mature male of a specific rank and an estrous female – was under observation either in

Table XXIV. Mean rates of copulation, all females combined; giving the pooled mean rates/h of copulation between all estrous females combined and sexually mature males of all ranks by cycle day as calculated from focal male and focal female samples of behavior by a method described in the text. Figure beneath rate is the time-base (in minutes) on which it was calculated. Normalized mean rates of copulation given here are plotted against rank of male in figures 20 and 21. Procedures for recovery of raw rank-specific frequency of copulation data are given in the text

Rank	Cycle day									
	D-7	D-6	D-5	D-4	D-3	D-3[1]	D-2	D-1	D-7 to D-1[2]	D-7 to D-1[1]
1	0.00	0.13	0.09	0.09	0.22	0.24	0.09	0.04	0.09	0.09
	1,489.00	1,841.33	2,088.75	2,077.30	2,421.55		2,688.15	2,689.88		
2	0.25	0.40	0.42	0.27	0.25	0.28	0.35	0.15	0.30	0.31
	1.458.75	1,811.74	1,998.75	2,196.79	2,387.60		2,718.18	2,719.88		
3	0.12	0.22	0.16	0.17	0.19	0.21	0.24	0.11	0.17	0.18
	1,489.00	1,901.74	1,908.75	2,167.30	2,510.63		2,748.97	2,629.88		
4	0.00	0.07	0.03	0.06	0.00	0.00	0.00	0.02	0.03	0.03
	1,459.00	1,781.74	2,028.75	2,167.30	2,472.15		2,617.42	2,479.88		
5	0.08	0.03	0.03	0.03	0.08	0.09	0.09	0.09	0.06	0.06
	1,429.00	1,721.74	1,968.35	2,167.30	2,361.44		2,714.09	2,659.88		
6	0.08	0.00	0.03	0.00	0.00	0.00	0.00	0.00	0.02	0.02
	1,458.40	1,751.74	1,908.32	2,055.90	2,391.55		2,627.84	2,629.88		
7	0.14	0.03	0.13	0.02	0.02	0.02	0.00	0.06	0.06	0.06
	1,699.00	2,201.74	2,358.75	2,467.30	2,661.55		2,808.97	2,959.05		
8	0.09	0.04	0.07	0.00	0.03		0.05	0.02	0.04	0.04
	1,278.68	1,539.59	1,758.75	1,927.30	2,360.11		2,538.97	2,625.42		
9	0.12	0.13	0.09	0.03	0.05	0.06	0.04	0.07	0.08	0.08
	1,517.21	1,838.39	2,024.79	2,227.30	2,421.55		2,748.97	2,629.88		
10	0.06	0.00	0.12	0.03	0.06	0.07	0.03	0.12	0.06	0.06
	1,049.00	1,200.00	1,528.38	1,783.99	1,994.20		1,923.61	2,017.48		
11	0.00	0.00	0.05	0.00	0.00	0.00	0.00	0.00	0.01	0.01
	729.88	952.56	1,325.05	1,400.14	1,564.23		1,560.81	1,548.58		
12	0.00	0.07	0.17	0.00	0.00	0.00	0.00	0.00	0.03	0.03
	729.88	847.56	1,050.94	1,103.21	1,299.28		1,362.74	1,358.58		
13	0.00	0.00	0.00	0.00	0.00	0.00	0.00	0.04	0.01	0.01
	729.88	847.56	1,050.94	1,048.56	1,239.28		1,302.74	1,358.58		
14	0.00	0.00	0.00	0.00	0.00	0.00	0.00	0.00	0.00	0.00
	499.88	537.56	791.91	747.73	800.59		754.18	725.58		
Mean	0.07	0.08	0.10	0.05	0.06		0.06	0.05	0.07	
Mean rank	5.45	4.07	5.71	3.20	3.47		3.39	5.74	4.50	

1 Mean rate, normalized
2 Mean rate

focal animal samples on the male or in focal animal samples on the female. The pooled mean rates of copulation were calculated according to the extension of formula 4 discussed above, and the raw rank-specific frequency of copulation data can be recovered from table XXIV by multiplying the rate in each cell of the table by 1/60th of the number of minutes shown in the table immediately beneath the rate. Examination of table XXIV shows that all ranks other than rank 14 had a nonzero pooled mean rate of copulation with estrous females on one or more of the cycle days analyzed. However, in general, second and third ranking males had a higher rate of copulation with females on the cycle days analyzed than did males of any other dominance rank. Thus, it may not be unrelated that second and third ranking males underwent more rank changes and more days of inconsistent dominance relationship than did males of any other rank (fig. 12) [cf. ROSE *et al.*, 1971].

Proportion of Copulations by Males of Each Rank

Figure 20 presents graphically the proportion, C_r, of observed copulations on cycle day D-3 in which the male partner was of rank r as well as the proportion for each rank expected under the priority-of-access model. These data are taken from tables XXIV and XXIII, respectively. Figure 21 presents comparable data on proportion of copulations based on the mean rate of copulation for each rank taken over cycle days D-7 through D-1 inclusive (table XXXVII). Cycle day D-3 is the day of maximum fertility and cycle days D-7 through D-1 are the period of potential fertile matings (cf. chapter III) [HENDRICKX and KRAEMER, 1969].

Since a maximum of three estrous females were present in the group on any day of study, only the first through third ranking males were expected to have a nonzero proportion of copulations attributable to them. In fact, on 85% of all study days (N = 400 days) there was either one estrous female or no estrous female present in the study group, so only a very low proportion of copulations were expected under the model for any male other than the first ranking individual. However, both figures 15 and 16 demonstrate that first ranking males had a far lower likelihood of copulating with an estrous female than was predicted by the model. The proportion of copulations carried out by males of the first three ranks differed significantly from expectations under the model (Chi^2 one sample test, X^2 = 229.36 and 251.44 for figures 20 and 21, respectively, d.f. = 2, 0.05 level of significance). Thus, the revised priority-of-access model did not adequately describe the mating patterns in this group of Amboseli baboons.

Cycles Resulting in Pregnancy

Three females, Oval, Ring, and Fluff, became pregnant during the study period, and though Ring and Fluff both gave birth to viable offspring, Oval's infant was stillborn. The exact number of minutes that each of these three females was under observation in *ad libitum* or focal female samples during each day of the estrus that resulted in pregnancy is shown in table XXV along with the number of copulations recorded in each sample record and the rank and identity of the male partner in each copulation. No focal male samples were being taken when these three females became pregnant.

Two males, Stubby and BJ, were alternately first ranking during the three cycles that resulted in pregnancies (fig. 9). Stubby, while first ranking was observed to copulate three times in one of the cycles that resulted in pregnancy; Stubby's copulation partner was female Oval. BJ, while first

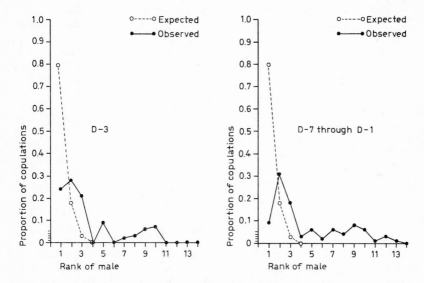

Fig. 20. Priority-of-access model, cycle day D-3; showing a test of the priority of access model with normalized rate of copulation data for cycle day D-3 as calculated from both focal male and focal female samples by a method described in the text. Raw frequency of copulation data can be recovered from table XXIV, on which this figure is based, by a method described in the text.

Fig. 21. Priority-of-access model, cycle days D-7 through D-1; showing a test of the priority-of-access model with normalized rate of copulation data for cycle days D-7 through D-1 inclusive as calculated from both focal male and focal-female samples of behavior by a method described in text. Raw frequency of copulation data can be recovered from table XXIV, on which this figure is based, by a method described in the text.

Table XXV. Copulation data for cycles resulting in pregnancy, giving the copulation partners and rates/h of copulation for estrous females by cycle day for the three cycles that resulted in pregnancy. Data from focal female and *ad libitum* samples as indicated; no focal male samples were obtained during these cycles

Cycle day	Sampling technique	Minutes of sample time	Copulation partner and rank		Number of copulations	Mean rate/h	Mean partner rank
A. Female Oval: 25–31 October 1971							
D-7	*Ad libitum*	566	none	–	–	0.00	–
	Focal	80.00	Ivan	2	1	1.50	4.0
			Sinister	6	1		
D-6	*Ad libitum*	213	none	–	–	0.00	–
	Focal	80.00	Dutch	5	1	0.75	5.0
D-5	*Ad libitum*	646	none	–	–	0.00	–
	Focal	78.05	Ivan	2	2	1.50	2.0
D-4	*Ad libitum*	628	Dutch	5	1	0.09	5.0
	Focal	80.00	Stubby	1	1	0.75	1.0
D-3	*Ad libitum*	228	none	–	–	0.00	–
	Focal	60.00	Ivan	2	1	2.00	6.0
			Even	10	1		–
D-2	*Ad libitum*	595	Stubby	1	1	0.10	1.0
	Focal	60.00	none	–	–	0.00	–
D-1	*Ad libitum*	507	Stubby	1	1	0.12	1.0
	Focal	80.00	none	–	–	0.00	–
Mean	*Ad libitum*					0.04	2.2
	Focal					0.93	3.9
B. Female Ring: 31 October–6 November 1971							
D-7	*Ad libitum*	507	none	–	–	0.00	–
	Focal	40.00	Ivan	2	1	1.50	2.0
D-6	*Ad libitum*	505	none	–	–	0.00	–
	Focal	80.00	Ivan	2	2	1.50	2.0
D-5	*Ad libitum*	531	Even	11	1	0.11	11.0
	Focal	80.00	none	–	–	0.00	–
D-4	*Ad libitum*	208	none	–	–	0.00	–
	Focal	77.98	none	–	–	0.00	–
D-3	*Ad libitum*	519	Dutch	5	1	0.23	7.0
	Focal	80.00	Max	9	1		
D-2	*Ad libitum*	526	Dutch	5	1	0.11	5.0
	Focal	40.00	Ivan	2	1	6.00	6.5
			Dutch	5	1		
			Max	9	1		
			Ben	10	1		
D-1	*Ad libitum*	583	none	–	–	0.00	–
	Focal	80.00	none	–	–	0.00	–
Mean	*Ad libitum*					0.06	7.5
	Focal					1.39	5.0

Table XXV (continued)

Cycle day	Sampling technique	Minutes of sample time	Copulation partner and rank		Number of copulations	Mean rate/h	Mean partner rank
C. Female Fluff: 16–22 December 1971							
D-7	*Ad libitum*	136	none	–	–	0.00	–
	Focal	30.00	none	–	–	0.00	–
D-6	*Ad libitum*	236	Stubby	2	1	0.25	2.0
	Focal	80.00	none	–	–	0.00	–
D-5	*Ad libitum*	92	none	–	–	0.00	–
	Focal	40.00	Stubby	2	2	3.00	2.0
D-4	*Ad libitum*	604	Stubby	2	2	0.19	2.0
	Focal	40.00	Stubby	2	2	3.00	2.0
D-3	*Ad libitum*	367	Stubby	2	1	0.16	2.0
	Focal	80.00	Stubby	2	1	0.75	2.0
D-2	*Ad libitum*	570	none	–	–	0.00	–
	Focal	80.00	Stubby	2	3	2.25	2.0
D-1	*Ad libitum*	196	none	–	–	0.00	–
	Focal	78.90	Stubby	2	1	0.76	2.0
Mean	*Ad libitum*					0.09	2.0
	Focal					1.39	2.0

ranking, was not observed to copulate with any of these females during the estrus that resulted in pregnancy. Thus, these limited observations, based both on *ad libitum* and focal female data, do not support the hypothesis that first ranking males have a higher reproductive success than do lower ranking males. In fact, it is clear from table XXV that no one male was observed to mate with both females Ring and Fluff, the two females who produced viable young, during the estrus that resulted in pregnancy and thus it seems likely that the two infants born into Alto's group during the study period each had a different father. In later sections of this chapter, aspects of male behavior that may have influenced reproductive success will be discussed; however, the present sample of menstrual cycles that resulted in conception is insufficient to determine if fertile cycles differed from nonfertile cycles in terms of length or other characteristics.

Correlation of Dominance Rank and Reproductive Success
Although the priority-of-access model did not adequately fit the observed pattern of mating in the study group, a positive correlation between dominance rank in males and the proportion of copulations, C_r, attributable

to males of each rank was found. The proportion of copulations on cycel day D-3 attributable to males of each rank (table XXIV) correlated positively with male dominance rank (Spearman rank correlation coefficient, r_s = 0.70; significant at 0.01 level). The correlation between dominance rank and proportion of copulations attributable to males of each rank over cycle days D-7 through D-1 inclusive was even stronger than that for cycle day D-3 alone (r_s = 0.75; 0.01 level of significance), though a correlation of 0.75 indicates that dominance rank accounted for only 56% of the variance in proportion of copulations. It should be emphasized, however, that in both figures 20 and 21, first through third ranking males accounted for over half of all copulations with estrous females.

Therefore, a relationship between dominance rank in males and reproductive success does exist, but the exact nature of this relationship remains to be determined. Whatever the exact relationship between dominance rank and reproductive success in males, it is clear that differential reproduction in baboons must depend on other factors in addition to dominance rank. In the sections that follow, selectivity of mating behavior and other important features of the mating system of Amboseli baboons will be discussed. Hopefully, this discussion will provide the information necessary for the development of more powerful models of baboon mating systems.

Mating Behavior in Alto's Group

Deviations from the Model. Selectivity in Mating Behavior

The primary deviation of the observed pattern of mating in the study group from that predicted by the priority-of-access model was that first ranking males did not consort and copulate with estrous females on every potentially fertile day. As will be shown below, this phenomenon resulted both from cycle day selectivity in mating and from selectivity in consort partner choice, or 'favoritism'. In terms of the priority-of-access model, selectivity of mating behavior meant that estrous females did not always consort or mate with the highest-ranking male partner available, a violation of one assumption of the model.

Cycle Day Selectivity in Copulatory Activity

Table XXIV lists the mean rank of the copulation partners of estrous females on cycle days D-7 through D-1. It will be seen that in general, the mean rank of copulation partners of estrous females was highest on cycle days D-4 through D-2, or in other words, higher ranking males concentrated their copulations around the optimal day for mating, cycle day D-3.

Figure 22 is a more detailed analysis of the distribution of copulatory activity by high ranking males. This figure presents the proportion of copulations by males of ranks one, two, and three, that were carried out on each cycle day. Only males of these three ranks had a nonzero expected proportion of matings under the model and in fact these males accounted for over half of all copulations with estrous females (tables XXIII, XXIV; fig. 21). In figure 22 it will be seen that males of ranks two and three carried out about 12–20% of their copulations on each of cycle days D-4 through D-2 or 42–50% of all their copulations on these 3 days combined. In contrast, first ranking males carried out about 33% of their copulations on D-3 alone, or 61% of all their copulations on cycle days D-4 through D-2 combined. The proportion of copulations carried out by first ranking males on cycle days D-4 through D-2 combined was significantly greater than the proportion of copulations carried out by second and third ranking males on these cycle days, at least as these proportions were estimated from the focal sample data (Chi^2 two-sample test, $\chi^2 = 8.78$; d.f. $= 1$; 0.01 level of significance). Thus, although figures 20 and 21 show that second and third-ranking males completed a larger proportion of all the copulations that occurred on cycle day D-3 than did the first ranking male, figure 22 demonstrates that a first ranking male was far more likely than males of other ranks to carry out his

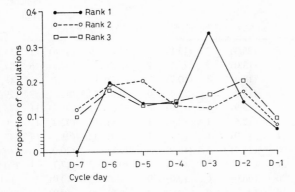

Fig. 22. Distribution of copulatory activity for high ranking adult males, showing the proportion of copulations by males of ranks 1–3 that were completed on each potentially fertile cycle day. See text for further explanation. Raw frequency of copulation data can be recovered from table XXIV, on which this figure is based, by a method described in the text.

copulations on cycle day D-3. The same conclusion applies to the distribution of male consort activity, as discussed below. This pattern of selective mating and consort formation is considered *cycle day selectivity* in mating behavior.

Cycle Day Selectivity in Consortship Formation

Tables XXVI and XXVII provide a detailed analysis of consortship formation by cycling females. These tables are based on data on the identity and rank of the consort partner of cycling females in each morning and afternoon focal female sample obtained on cycle days D-7 through D+3 and presented in HAUSFATER [1974]. On days when a female was focal sampled, two samples were taken: one in the morning and one in the afternoon of the same day. For each cycle day, table XXVI gives the proportion of sample days on which a female had a consort in neither morning or afternoon sample, in the morning sample only, in the afternoon sample only, or in both samples. Examination of table XXVI indicates that on cycle days

Table XXVI. Cycle day selectivity in consortship formation, based on data in HAUSFATER [1974]. See text for explanation

Cycle day	(A) Number of cycles with both morning and afternoon focal samples	(B) No consort morning or afternoon % of (A)		(C) Consort in morning only % of (A)		(D) Consort in afternonn only % of (A)		(E) Consort both morning and afternoon % of (A)		(F) Morning consort rank ≠ afternoon consort rank % of (E)	
D-7	19	5	(26.3)	3	(15.8)	2	(10.5)	9	(47.4)	5	(55.6)
D-6	23	3	(13.0)	2	(8.7)	2	(8.7)	16	(69.6)	6	(37.5)
D-5	28	3	(10.7)	3	(10.7)	5	(17.9)	17	(60.7)	7	(41.2)
D-4	29	4	(13.8)	1	(3.4)	0	(0.0)	24	(82.8)	7	(29.2)
D-3	34	2	(5.9)	1	(2.9)	0	(0.0)	31	(91.2)	11	(35.5)
D-2	38	1	(2.6)	1	(2.6)	1	(2.6)	35	(92.2)	13	(37.1)
D-1	35	1	(2.9)	1	(2.9)	1	(2.9)	32	(91.3)	14	(43.7)
D	30	8	(26.7)	5	(16.7)	5	(16.7)	6	(20.0)	0	(0.0)
D+1	26	22	(84.6)	2	(7.7)	0	(0.0)	2	(7.7)	1	(50.0)
D+2	20	20	(100.0)	0	(0.0)	0	(0.0)	0	(0.0)	–	(–)
D+3	14	14	(100.0)	0	(0.0)	0	(0.0)	0	(0.0)	–	(–)
Mean D-7EMD-1	29.4	2.7	(10.7)	1.7	(6.7)	1.6	(6.1)	23.4	(76.5)	9.0	(39.97)
Mean D-7EM D+3	25.6	7.6	(35.1)	2.3	(0.3)	1.5	(5.4)	15.6	(51.2)	7.1	(36.64)

D-7 through D-5 females lacked a consort in one or both samples on the same day in one half of all cycles. In contrast, on cycle days D-4 through D-1, females had a consort in *both* the morning and the afternoon samples of the same day in about 90% of all cycles. However, table XXVI also indicates that on about one third of all cycle days that females did have both a morning and afternoon consort on the same day, the identity, and therefore rank, of the afternoon consort was different from that of the morning consort. Thus, the fact that a specific female was in consort for an entire day does not imply that her consort partner remained the same throughout the day. On the day of deturgescence, females were most likely either to lack a consort altogether or to have a consort only during the morning sample. Finally, on cycle day D + 1, females only rarely had a consort in either sample and no female was ever observed to consort on cycle days D + 2 or D + 3.

Thus, it is clear that males did not consort with a female over her entire period of fertile matings, but that consortships were formed selectively and were most frequent on the optimal cycle days for mating.

Rank of Male Consort Partners

Table XXVII presents information on the mean rank of the consorts of estrous females in the morning and afternoon focal female samples on

Table XXVII. Daily changes in consort partner, giving the mean morning, afternoon, and daily rank of consort partners of estrous females by cycle day, based on data in HAUSFATER [1974]

	Cycle day										
	D-7	D-6	D-5	D-4	D-3	D-2	D-1	D	D+1	D+2	D+3
Mean rank of consorts in morning samples	3.07	2.37	2.71	2.52	1.94	2.00	1.95	3.68	4.25	–	–
Number of cycles	15	19	21	29	33	38	36	19	4	–	–
Mean rank of consorts in afternoon samples	4.18	2.89	2.77	2.65	2.91	2.28	2.97	3.36	4.50	–	–
Number of cycles	11	19	22	26	33	36	36	11	2	–	–
Mean morning and afternoon consort ranks	3.63	2.63	2.74	2.59	2.43	2.14	2.46	3.52	4.38	–	–
Mean afternoon rank minus mean morning rank	1.11	0.52	0.07	0.13	0.97	0.28	1.02	en0.32	0.25	–	–

cycle days D–7 through D + 3. This table also includes the mean daily rank of consorts on each cycle day calculated as the mean of the morning and afternoon mean consort ranks. In general, the mean daily rank of consorts of estrous females did not change greatly over the period of fertile mating, always remaining between two and three, with the exception of cycle day D-7. Thus, the selectivity of consortship formation discussed above was apparently practiced by males of all ranks, not just high ranking males. However, it was also true that in general the morning consorts of estrous females were higher ranking than were the afternoon consorts. Considering only cycle days D-7 through D + 1, this difference is statistically significant (Wilcoxon matched-pairs signed-ranks test, 0.05 level of significance). This effect is in part accounted for by the morning-to-afternoon change of consort partners on those cycle days that females had a consort in both the morning and afternoon sample of the same day (see above). However, other factors must also be at work and it is tempting to speculate that the time of ovulation in baboons, like time of birth in man [JOLLY, 1972a], is not randomly distributed throughout the day [cf. BALIN and WAN, 1968]. Nevertheless, the data in tables XXVI and XXVII indicate that selectivity of male-female consortship formation was taking place both with respect to cycle day and time of day.

Individual Selectivity in Consort Partner Choice

Table XXVIII summarizes an analysis of selectivity of consort partner choice on cycle day D-3 as seen from the viewpoint of first ranking males; as shown in figure 22, males when ranked first completed most of their copulations on cycle day D-3.

Both first ranking males, BJ and Stubby, consorted with one or two females on nearly all of those females' D-3 cycle days and only rarely consorted on cycle day D-3 with any of the remaining three or four females. While Stubby was first ranking, five different females were sampled on cycle day D-3 in a total of 13 menstrual cycles. Stubby was the consort of female Oval in at least one sample on each D-3 cycle day in her three menstrual cycles. In the remaining ten menstrual cycles, Stubby consorted on cycle day D-3 only once, with female Judy. Similarly, while BJ was first ranking, five females were sampled on cycle day D-3 in a total of 25 menstrual cycles. BJ consorted with female Lulu in at least one sample on cycle day D-3 in eight (80%) of her ten cycles. BJ also consorted with female Judy in at least one sample on cycle day D-3 in five (71.4%) of her seven cycles. In the remaining eight cycles of three other females, BJ consorted on cycle day

Table XXVIII. Selectivity of consort partners, giving an analysis of selectivity of consort partner choice for first-ranking males on cycle day D-3, based on data in HAUSFATER [1974]. See text for explanation

Female	Number of cycles	Consortship with Stubby on cycle day D-3	%
A. Adult male Stubby			
Oval	3	3	100
Fluff	3	0	0
Ring	3	0	0
Judy	2	1	50
Lulu	2	0	0
Alto	–	–	–

Female	Number of cycles	Consortship with BJ on cycle day D-3	%
B. Adult male BJ			
Lulu	10	8	80.0
Judy	7	5	71.4
Alto	4	0	0
Oval	3	1	33.3
Fluff	1	0	0
Ring	–	–	en

D-3 only once with female Oval. Conversely, when females Alto, Ring, or Fluff were in estrus, first ranking males BJ and Stubby did not form consortships with them even if no other estrous females were present in the group. Thus, the observed pattern of selectivity of consort formation on cycle day D-3 cannot be explained solely by the distribution of multiple estrous females in the group. The three 'favorites' of the first ranking males, namely females Lulu, Oval, and Judy, all had the traits of light or gold colored fur, stocky physique, and all appeared relatively young compared to the other cycling females. Lulu and Judy both cycled throughout the study period and Oval underwent the next highest number of menstrual cycles.

While all of the above data on selectivity of mating and consortship formation have been treated as examples of male selectivity, they are equally well described as selectivity by estrous females. The relative contribution of the male and the female to the formation and maintenance of a consortship remains to be determined. In fact, as noted in chapter I, the theory of sexual selection actually predicts that female mammals will show strong mate preferences or selectivity and that male mammals will mate indiscriminately. Nevertheless, it should be clear from the above discussion that one major factor contributing to the observed deviations from predictions based on the priority-of-access model is selectivity in consortship formation and copulatory activity. In general, the net result of selectivity of consortship formation and copulatory activity was to enhance the mating frequency and presumably reproductive success of lower ranking males.

Male-Male Harassment and the Consort Relationship

DeVore [1965] and Hall and DeVore [1965] have given an extensive description of patterns of male-male harassment that occurred in two groups of anubis baboons *(Papio anubis)* in Nairobi National Park, Kenya. DeVore observed that some males, who as individuals were relatively low ranking, banded together and harassed and attacked the higher ranking male consort of an estrous female. Low ranking males often gained access to estrous females through such cooperative harassment of individually higher ranking males. Additionally, consorting males were sometimes prevented from completing their attempts at copulation by persistent harassment and physical interruption from other males in the group, though high ranking males nonetheless completed a larger proportion of their copulation attempts than did lower ranking males.

In general, my observations on the context and pattern of male-male harassment agree with DeVore's description. However, no significant difference in the proportion of mountings completed to ejaculation by sexually mature males of different ranks was found in the Amboseli study group. Table XXIX presents the number of complete and incomplete mountings by males of each rank as recorded in the focal male and focal female samples. Males of ranks 1–5 were grouped as high-ranking, males of ranks 6–9 as middle-ranking, and males of ranks 10–14 as low-ranking. No significant difference was found in the proportion of copulations completed by high-, middle-, and low-ranking males (Chi^2 one sample test $X^2 = 2.96$; d.f. $= 2$; 0.05 level of significance).

Actual physical interruption of a copulation was only rarely observed and only four instances of physical interruption of a copulation attempt were recorded in the focal animal samples. In two of these interruptions, a low ranking male ran screaming toward a higher ranking male engaged in copulation. In the other two instances, a high ranking male ran and batted at a low ranking male engaged in copulation. In all four cases, the interrupter was by himself rather than acting in coalition with other males.

Harassment and Subadult Male Behavior

A frequent component of harassment of a consort male was the counter-chase pattern (table V B). In a counter-chase, a subordinate male ran cackling, grimacing, and with tail up, toward a more dominant individual. The more dominant individual then loped away over a variable distance until the subordinate ceased his chasing, though often the counter-chase was terminated when the more dominant individual turned and physically attacked the subordinate. Nevertheless, a high ranking consort male was effectively forced away from the estrous female for a few minutes or longer by the counter-chase of lower ranking males, either singly or in pairs. Repeated counter-chasing often preceded the morning-to-afternoon change of consort partners, and counter-chasing may have lowered the reproductive success of higher ranking males.

Table XXIX. Proportion of copulation attempts completed, giving the proportion of complete and incomplete mountings by sexually mature males of each rank on estrous females by cycle day of female. A mounting was complete if male gave an ejaculatory pause or female had fresh ejaculate on her perineum immediately after mounting. A completed mounting is equivalent to a copulation as the term is used in this work

Rank	D-7 complete	D-7 incomplete	D-6 complete	D-6 incomplete	D-5 complete	D-5 incomplete	D-4 complete	D-4 incomplete	D-3 complete	D-3 incomplete	D-2 complete	D-2 incomplete	D-1 complete	D-1 incomplete	Total complete	Total incomplete	Complete, %
1	0	0	4	1	3	2	3	0	8	1	4	0	2	7	24	11	68.6
2	6	1	10	3	16	5	10	4	10	5	16	6	7	2	75	26	74.2
3	3	0	7	1	5	0	6	0	8	2	11	3	5	4	45	10	77.8
4	0	2	2	0	1	0	2	0	0	0	0	0	1	1	6	3	66.7
5	2	0	1	1	1	0	1	2	3	1	4	1	4	1	16	6	72.7
6	2	2	0	0	1	0	0	0	0	0	0	0	0	0	3	2	60.0
7	4	2	1	2	5	1	1	0	1	0	0	0	2	0	14	5	73.7
8	2	0	1	0	2	1	0	0	1	0	1	1	2	1	9	3	75.0
9	3	0	4	0	3	2	1	1	1	0	4	1	3	0	19	4	82.6
10	1	0	0	0	4	0	1	0	1	1	1	1	4	0	12	2	85.7
11	0	0	0	0	1	0	0	0	0	0	0	0	0	0	1	0	100.0
12	0	1	1	0	3	1	0	0	0	0	0	0	0	0	4	2	66.7
13	0	0	0	0	0	0	0	0	0	0	0	0	1	0	1	0	100.0
14	0	0	0	0	0	1	0	0	0	0	0	0	0	0	0	1	0.0
Total	23	8	31	8	45	13	25	7	33	10	41	13	31	16	229	75	
Complete, %	74.2		79.5		77.6		78.1		76.7		75.9		65.9		75.3		

However, harassment behavior was not directed only from lower ranking males to higher ranking males. In many afternoons the consort of an
estrous female was a middle or low ranking male. Often, a higher ranking
male persistently followed behind the consort pair at a distance of 50 ft or
more, and usually this was the same male who had been the female's morning
consort. The lower ranking afternoon consort repeatedly ran at the follower
in brief counter-chase and then quickly returned to the estrous female. It was
my impression that the presence of the higher ranking male inhibited copulation by the lower ranking consort male.

Subadult males, too, frequently trailed after a consort pair, but subadults
did so in a much less overt manner than did adult males. When the consorting
male was temporarily absent from the estrous female, as when harassment or
counter-chasing was in progress, the subadults quickly ran toward the estrous
females and attempted copulation (fig. 6). The subadults were notably successful in completing these copulations (table XXIX, ranks 8 and 9), but
only rarely did they receive overt aggression from the consort male on his
return. The enhanced mating success of subadult males (ranks 9 and 10)
compared to that predicted on the basis of their dominance rank is evident
in figures 20–21. Juvenile-2 males only rarely engaged in this particular
pattern of mating behavior and it is possible that this kind of behavior is
restricted to males in the subadult stage of their life cycle.

Discussion

What Determines Mating Patterns in a Baboon Group?

Although the priority-of-access model did not adequately describe the
mating patterns in Alto's group, it was nevertheless true that a strong positive
correlation between dominance rank and reproductive success in males was
found. However, a high correlation coefficient in itself does not explain the
mating patterns in the study group. In particular, it is necessary to explain
the fact that first ranking males carried out a far smaller proportion of all
copulations than was expected under the priority-of-access model.

The present study has established that first ranking males did *not* consort
with a particular subset of estrous females even when no other estrous females
were present in the group. Additionally, it has been shown that even when
first ranking males did consort and copulate with a female, they did *not* do
so on all of her potentially fertile cycle days. Instead, first ranking males
copulated and consorted most frequently on cycle day D-3, the day that

laboratory studies have shown to be the optimal time for mating in baboons. These data indicate that selectivity in consortship formation and copulation partners is one of the most important factors for understanding the mating patterns in baboons. A sexual pheromone similar to that found in rhesus monkeys [MICHAEL and KEVERNE, 1968] is one mechanism through which cycle day and individual selectivity may be mediated. Thus, further studies on reproduction in baboons should concentrate not on first ranking male behavior *per se*, but more generally on the behavioral and pheromonal mechanism of consortship formation and copulation partner choice.

The subset of estrous females that did not form consortships with the first ranking male consorted with the second, third, or fifth ranking male in the group instead. Although some individual selectivity or favoritism may have influenced these partner choices also, it was usually the case that a female did consort with the highest ranking of these males available to her. In effect, it was this aspect of mating behavior that produced the strong positive correlation between dominance rank and reproductive success in males in the study group. In other cases, females on potentially fertile cycle days did not form any consortship at all, but completed copulations outside of the consort relationship. As shown above, it was of course more common for a female to lack a consort on cycle days D-7 through D-5 than on cycle days closer to D-3. However, since females often lacked consorts on cycle days D-7 through D-5, and since they continued to copulate when not in consort, it is an open question as to what advantage, if any, was gained by males who did choose to consort on cycle days D-7 through D-5.

In sum, the mating system of Amboseli baboons may be viewed as including several alternative reproductive strategies for males. In any such short-term reproductive strategy, a male's potential gain in reproductive success from consorting with a female on a cycle day with only a low probability of conception must be weighed against his chances for gaining access to that female on the optimal day for mating and against whatever immediate or long-term negative consequences he may suffer for participating in a consortship regardless of cycle day. The male's chances of gaining access to the female on or near the optimal day for mating are probably strongly influenced by his dominance rank, while negative consequences of participating in a consortship may include reduced food intake, due to the need to constantly attend to the female, and increased exposure to aggression and harassment from other males.

The strategy characteristic of first ranking adult males is to mate and consort on only a few cycle days and with only a few particular individual females. However, this mating activity is concentrated around the optimal day for mating, cycle day D-3. Through this strategy, first ranking males have a relatively high probability of impregnating a female, but suffer minimal short-term negative consequences of consortship. The ability of first ranking males to be so selective is of course dependent upon their dominance rank. Other adult males exhibit a strategy of mating and consorting on as many potentially fertile cycle days as possible, including those cycle days more distant from the optimal day. A male's probability of access to estrous females on these cycle days is probably strongly influenced by his dominance rank, but consort formation on such days increases the male's exposure to whatever negative consequences of consorting may exist. Finally, subadult males exhibit elements of both strategies in their mating behavior: they selectively mate on the optimal cycle day, but are able to do so only by maintaining constant vigilance for the temporary absence of an estrous female's adult male consort. Although subadults do not consort, and thus do not suffer the negative consequences of consortship *per se*, their need for constant vigilance and special mating patterns probably produces marked changes in their food intake and activity patterns as well as continually exposes them to aggression from adult males. Obviously, a male would utilize a succession of these short-term strategies in his life time, depending on his age and dominance rank; further thoughts on this subject are presented in chapter VI.

Summary

The following is a summary of the most important points of this chapter:

1. A priority-of-access model of mating behavior was described. This model is composed of a small number of assumptions that, taken together, allow the use of the cumulative binomial probability distribution to predict the amount of time that males of each dominance rank will be in consort with estrous females.

2. The amount of time spent in consort with estrous females was judged to be a poor measure of reproductive success in baboons and instead the normalized frequency of copulation per unit time between males of each rank and estrous females was chosen as a measure of reproductive success for use in testing the model. The model was suitably modified by addition

of the assumption that males of rank r copulate with estrous females in proportion to the time that at least r females are simultaneously in estrus.

3. The revised priority of access model did not provide an adequate fit to the observational data on baboon mating behavior. In particular, first ranking males carried out a far lower proportion of copulations than predicted, and males of all ranks lower than three carried out a larger proportion of copulations than predicted. Second ranking males carried out the highest proportion of copulations during the study period.

4. The single most important factor in producing deviations from the model was that first ranking males copulated and consorted selectively with respect both to cycle day and time of day, rather than utilizing every potentially fertile cycle day. Additionally, first ranking males did not form consortships with a specific subset of individual estrous females even when no other estrous females were available. The net result of this selectivity of mating behavior was to enhance the mating success of lower ranking males.

5. Male-male harassment was described, but was probably of little importance in influencing the mating success of adult male baboons: males of all ranks completed about the same proportion of their attempts at copulation and instances of physical interruption of a copulation attempt were rare. Subadult males, however, successfully carried out copulations during temporary absences of an estrous female's adult male consort and such absences were usually caused by harassment of the consort by other adult males.

6. The mating system of Amboseli baboons was described in terms of several alternative short-term reproductive strategies shown by males.

V. Estrous Females and Group Organization

Introduction

This chapter is an analysis of the effects of the presence of estrous females on several aspects of social organization in the study group. Specifically, changes in (1) group composition, (2) dominance orderings, (3) rates of adult male social behavior, and (4) rates of wounding will be examined in relationship to changes in the number of estrous females in the group.

The question of changes in group organization as a result of the presence of an estrous female has direct relevance to a number of biological, psychological, and anthropological problems. On the one hand, estrus may be viewed as a hormonally induced change in social behavior, and thus this chapter and chapter III can be regarded as dealing with the hormonal basis of individual and group behavior. On the other hand, the presence of a female attractive to, and attracted by, males may be thought of as a social phenomenon that produces marked changes in the integration of the group. The presence of an estrous female in a social group of baboons has been called both a unifying [ZUCKERMAN, 1932] and a disruptive [WASHBURN and DEVORE, 1961a, b] social factor. However, the hormonal and social aspects of estrous behavior are neither unidirectional in effect nor unrelated. Thus, while patterns of social interaction for females may change as a result of their hormonal state, the converse is also true: SMITH *et al.* [1967], ROWELL [1970], and HENDRICKX and KRAEMER [1971] have shown that the social condition in which captive baboon females are maintained can influence both cycle length and fertility.

Group Changes and Changes in Dominance Relationships

Dominance Relations of Estrous Females
According to HALL and DEVORE [1965], the dominance rank of females in groups of anubis *(Papio anubis)* and chacma *(Papio ursinus)* baboons

fluctuated as a result of sexual cycling. In particular, estrous females were said to gain temporarily a dominance rank equivalent to that of their male consort, though no data in support of this hypothesis have ever been published.

Table VII summarizes data from the present study on the outcome of decided agonistic bouts of females on days D-7 through D + 3 of their menstrual cycles. As shown above, females usually did not consort on every day of this period, and females were particularly unlikely to consort on cycle days D + 2 and D + 3 (table XXVI). Nevertheless, this 11-day period (D-7 through D + 3) will be used in the following discussion as a first approximation to the period of consortship and estrus (i.e. cycle days D-7 through D-1 inclusive).

For each pair of females, the directionality and outcome of the majority of their decided agonistic bouts and the resulting rank ordering of the females was the same during their estrus focal samples (table VII) as in their *ad libitum* samples, which were taken primarily during nonestrus times (table VI). (Some portion of the *ad libitum* data doubtless came from times when the females were in consort, but by far the *ad libitum* samples gave most extensive coverage to nonestrus portions of the menstrual cycle.) If the dominance rank of females were to change as a result of their cycle state or consort partner, one would expect numerous reversal outcomes during estrus, i.e. data below the main diagonal of table VII. In fact this was not the case; rather, the available data indicate that the rank ordering of adult females was not affected by estrus or consort relationships.

Inconsistent Female Dominance Relationships

One period of inconsistent dominance relationship between adult females resulted from the single reversal in the Preg-Scar dyad (table VI). At the time of the reversal, female Scar, the winner, had a maximally swollen sexual skin and was in consortship with male Ivan, who was then first ranking male in the group. However, one day later Preg defeated Scar in an agonistic bout, though Scar was still swollen and in consort with Ivan. Hence, the previous defeat of Preg by Scar was not obviously the result of Scar's consortship with Ivan or of her swollen condition.

There were only a few other inconsistent dominance relationships between females, but none was clearly the result of consortship or estrus. For example, the relationships between J2 female Fem and adult female Judy, and between Fem and adult female Ring showed numerous periods of inconsistency in dominance relationship. However, only one of 22 decided agonistic bouts within these two pairs of animals occurred while either adult female was on cycle day D-7 through D+3 (table VII). The inconsistency in dominance relationship between adult females and maturing males (table IX) included both females who cycled during the study period and females who did not cycle. Therefore, the present data do not support the hypothesis that a female's dominance rank changed during estrus or as a result of her consort partner's rank.

Although in the present study no evidence was found that a female's dominance rank changed as a result of her cycle state, figure 15 shows that females' rates of participation in agonistic bouts did change in relation to cycle day. Briefly, the rate at which females won and lost agonistic bouts, exclusive of herding by mature males, appears to have been suppressed during estrus compared to other cycle states. In contrast, the rate of herding of females by mature males, which was negligible during other cycle states, increased greatly during estrus. Thus, while female dominance, defined as the number of other individuals who were consistently defeated in agonistic bouts, did not change as a result of estrus, the number and kinds of agonistic bouts in which a female was a participant did change as a result of estrus.

Group Changes by Adult Females

A total of four short-term group changes by females were recorded during the present study (table III). As noted previously, it seems likely that all short-term changes that occurred during the 400-day study period were recorded. Adult female Ring was absent from Alto's group from 31 October 1971 through 2 November 1971 and again from 28 February 1972 through 2 March 1972. During this second absence, Ring was accompanied by J1 female Gin, presumed to be Ring's daughter. Ring and Gin were with High Tail's group for the entire duration of both absences from Alto's group. Finally, J2 female Slinky, regularly a member of High Tail's group, was with Alto's group on 21–25 February 1972. High Tail's group was completely out of visual and vocal contact with Alto's group during most of Ring's stays with High Tail's group and during most of Slinky's stay with Alto's group, though the two groups did meet briefly at times.

Each of these intergroup movements of females occurred in the morning after the two groups had spent the previous night in adjacent sleeping groves, a common occurrence in the Reserve. The groups descended from the trees in the morning, remained beneath the grove briefly, and then dispersed for their daily foraging. As they did so, the migrant female casually moved away with the group other than that with which she was usually associated.

The females traveling with a different group did not attempt to return to their regular group at the first opportunity. Thus, Alto's group foraged near High Tail's group on one occasion while Slinky was with Alto's group. Slinky did not approach or interact with any of the individuals in High Tail's group although less than 300 ft separated the two groups for a short time. Ring, who left two presumed offspring in Alto's group during one of her absences, did not give calls typical of a mother who was separated from her infant, nor did her presumed younger daughter, Gin, give a 'lost infant' vocalization. Therefore, the short-term group changes of females probably were not the result of a female simply becoming confused and mistakenly following the wrong group away from the sleeping groves in the morning.

Ring had a small sexual skin swelling during her first short-term group change, but she was neither in consort nor maximally swollen. Ring was pregnant at the time of her second absence from the group. Slinky had a small sexual skin swelling characteristic of older juvenile females during her stay with Alto's group. Slinky was mounted once without ejaculation by a J2 male, Red, but otherwise showed no sexual activity. Thus, the available evidence does not indicate that group changes by females were related to the female's reproductive condition. A more detailed discussion of intergroup relations will be published elsewhere.

Group Changes by Adult Males

A number of plausible hypotheses concerning the relationship between the presence of estrous females in a social group and changes in the adult male composition of that group have been proposed [ZUCKERMAN, 1931, 1932; WASHBURN and HAMBURG, 1965]. For example, competition by adult males for access to estrous females may produce increased aggression in the group and result in some adult males leaving the group or being killed in fights. Alternatively, adult males may be strongly attracted by estrous females, and thus more likely to remain with the group or to join it from outside when estrous females are present.

Table III lists a total of 30 changes in the adult male composition of Alto's group during the study period, and as with the adult females it is believed that all such changes that occurred during the study period were recorded. These changes in male group membership varied from a brief one-day absence to more lengthy changes of a month or longer. Fifteen of these changes were departures, i. e. emigrations or death, and an equal number were immigrations. The exact date of all 15 immigrations is known, but the dates of two departures from the group, those of Peter and Sinister in late December, 1971, are not known precisely. No males matured into the adult class

Table XXX. Relationship between group changes by adult males and number of estrous females in group. Listed are changes in the adult male composition of the group by number of estrous females in group when change occurred. Estrus is defined as cycle days D-7 through D-1

Number of estrous females in group	Number of study days (%)		Number of immigra- tions of known date (%)		Num- ber expec- ted	Number of departures of known date (%)		Num- ber expec- ted	Number of durative immigra- tions (%)		Number of dura- tive de- partures (%)	
0	158	(39.5)	6	(40.0)	5.9	5	(38.5)	5.1	2	(40.0)	2	(40.0)
1	188	(47.0)	5	(33.3)	7.1	5	(38.5)	6.1	2	(40.0)	2	(40.0)
2	46	(11.5)	4	(26.7)	1.7	2	(15.4)	1.5	1	(20.0)	1	(20.0)
3	8	(2.0)	0	(0)	0.3	1	(7.6)	0.3	0	(0)	0	(0)
Total	400	(100.0)	15	(100.0)	15.0	13	(100.0)	13.0	5	(100.0)	5	(100.0)

during the study period, and thus the number of adult males in the group at the end of the study was the same as the number present at the beginning of the study.

Table XXX lists the number of all departures of known date and all immigrations of known date that occurred on days when no females were in estrus and on days when from one to three females in the group were in estrus (cycle days D-7 through D-1). On 47.0% of all study days (N = 400 days) only one female in the group was in estrus, and thus, 47.0% of all departures and 47.0% of all immigrations would be expected to have occurred on these days, and similarly for other numbers of estrus females. In fact, the proportion of all departures of known date, and the proportion of all immigrations of known date that occurred on days with 0, 1, 2, and 3 estrous females present in the group did not differ significantly from the proportions expected from the number of such days during the study period (Chi^2 one sample test, $\chi^2 = 1.00$ and $\chi^2 = 4.62$ for departures and immigrations, respectively; d.f. = 2; 0.05 level of significance). Examination of the 10 changes in male group membership that lasted for one month or longer showed the same results (tables III, XXX).

Changes in Dominance Rank Among Adult Males

Speculations have been offered that in a primate group changes in male dominance relationships may be related to the presence of estrous females in the group [WASHBURN and HAMBURG, 1965]. In essence, competition for access to estrous females is believed to increase aggression between adult males, and it is further speculated that increased aggression may result in a change in male dominance relationships. In fact, CHANCE [1956] reported that in a recently formed colony of rhesus monkeys *(Macaca mulatta)* aggression between adult males and changes in their dominance relationships seemed to be correlated with the onset of estrus in females in the colony. In the present study, an anylysis of the effects of the presence of estrous females on adult male dominance relationships was carried out in the same manner as the analysis of group changes by adult males.

Table XIV shows that on 85 study days *at least* one pair of adult males was inconsistent in their dominance relationship. Not all inconsistencies in dominance relationships resulted in a rank change, but by definition, all rank changes were preceded by a period of inconsistent dominance relationship. The number of days of inconsistent dominance relationship that occurred when none of the females in the group was in estrus and when one or more females in the group was in estrus is shown in table XXXI. The observed proportion of days of inconsistent dominance relationships among adult males that occurred when various numbers of females in the group were in estrus (cycle days D-7 through D-1) differed significantly from expectations based on the observed distribution of number of females in estrus (Chi^2 one-sample test, $\chi^2 = 9.88$; d.f. = 3; 0.05 level of significance). In particular, inconsistencies in the dominance relationships of adult males occurred less frequently than expected on days when none of the females in the group was in estrus, and more frequently than expected on days when

Table XXXI. Relationship between inconsistent dominance relationships in adult males and number of estrous females in group. Given is the number of days that *at least* one pair of adult males was inconsistent in dominance relationship by number of estrous females in group. Estrus is defined as cycle days D-7 through D-1

Number of estrous females in group	Number of study days		Number of days with *at least* one pair of adult males inconsistent in dominance relationship	Number expected
		(%)	(%)	
0	158	(39.5)	20 (23.5)	33.6
1	188	(47.0)	52 (61.2)	40.0
2	46	(11.5)	12 (14.1)	9.7
3	8	(2.0)	1 (1.2)	1.7
Total	400	(100.0)	85 (100.0)	85.0

one female in the group was in estrus (table XXXI). Days with two or three estrous females made minor contributions to the total χ^2 value. Thus there is evidence in the present study that inconsistencies in the dominance relationships of adult males, and hence rank changes, were related to the presence of estrous females in the study group.

Rates of Social Interaction by Adult Males

Rate Calculation

Table XXXII lists the number of minutes that each adult male in Alto's group was under observation in focal male samples of behavior. The focal-male samples and behaviors recorded in these samples were described in detail in chapter I. The minutes of observation for each male are separated according to the number of estrous females in the group when the sample was obtained. Although over half of all sample minutes were obtained on days when at least one female in the group was in estrus (cycle days D-7 through D-1), less than 4% of all sample minutes were obtained when more than one female was in estrus ($N = 10{,}518.71$ min).

Table XXXII. Focal male sample times, listing the number of minutes that individual adult males were under observation in focal male samples of behavior by number of estrous females present in group when sample was obtained[1]

Male	Number of estrous females in group				
	0	1	2	3	Total
BJ	659.69	689.59	30.00	30.00	1,409.28
Peter	647.48	594.50	30.00	0.00	1,271.98
Ivan	539.34	480.00	60.00	0.00	1,079.34
Crest	47.00	239.08	0.00	30.00	316.08
Stubby	686.67	717.85	20.60	30.00	1,455.12
Dutch	590.63	714.80	30.00	0.00	1,335.43
Sinister	708.37	772.49	59.30	0.00	1,540.16
Max	712.12	748.57	60.00	0.00	1,520.69
Cowlick	146.85	443.78	0.00	0.00	590.63
Total	4,738.15	5,400.66	289.90	90.00	10 518.71

1 Multiplication of the rates of male behavior shown in figures 23 and 24 by 1/60th of the number of minutes shown in the 'total' row will result in recovery of raw frequency data from all focal male samples combined.

In calculating rates of male behaviors, information about the behavior of each adult male was taken from focal samples on that male only. The rate of participation by each adult male in several categories of social interaction and the rate of masturbation by each adult male when none of the females in the group was in estrus, when one female, and when more than one female, was in estrus were calculated by formula 1 and are given in HAUS-FATER [1974] as are procedures for recovery of the raw frequency data from focal male samples.

As in the analysis of rates of social interaction of estrous females (chapter III), a mean rate of participation in each category of social interaction for all adult males was calculated by two different methods: a mean male rate (formula 2) and a pooled mean rate (formula 3) for all adult males combined. Table XXXIII gives the mean male rate of participation in each category of social interaction as well as the standard error of the mean male rate and one standard deviation. The category 'all nonagonistic interactions' is based on all occurrences of presentation, mounting, or grooming involv-

Table XXXIII. Mean male rates of social interaction, giving the mean rates (per male/h) of participation by adult males in various categories of social interaction by number of estrous females in group as calculated by the method described in the text

Category of interaction	No estrous females			One estrous female			> 1 estrous female		
	mean	SE	SD	mean	SE	SD	mean	SE	SD
Agonistic interactions									
Adult male-adult male decided agonistic bouts	3.03	0.79	2.38	2.13	0.26	0.78	1.15	0.40	1.12
Adult male-other than adult male decided agonistic bouts	3.20	0.88	2.64	2.62	0.66	1.97	2.80	1.11	3.14
Undecided agonistic bouts	0.21	0.14	0.42	0.13	0.07	0.22	0.00	0.00	0.00
Multi-individual agonistic bouts	0.04	0.02	0.07	0.10	0.05	0.15	0.00	0.00	0.00
All agonistic interactions	6.48	1.75	5.26	4.98	0.90	2.69	3.95	1.13	3.21
All nonagonistic interactions (presentation, mount, groom)	3.22	1.37	4.11	1.98	0.36	1.08	3.30	1.73	4.88
All social interactions	9.70	3.07	9.22	6.96	1.21	3.64	7.25	1.70	4.81
Masturbation	0.04	0.03	0.08	0.14	0.05	0.16	0.25	0.25	0.61

ing the adult male as either the actor or recipient of the behavior. The pooled mean rates of participation by adult males in these same categories of social interaction are presented graphically in figures 23 and 24, and will be discussed in the remainder of this chapter. However, examination of table XXXIII indicates that the mean male rates of agonistic interaction decreased as the number of estrous females in the group increased; and this inverse relationship is opposite to that indicated by HALL [1962] for chacma baboons *(Papio ursinus)*. In contrast, the mean male rates of nonagonistic interaction showed no simple relationship to the number of estrous females in the group.

Estrous Females and Changes in the Rates of Social Interaction Among Adult Males

Figure 23 shows the pooled mean rates of participation by adult males in all categories of social interaction ('all social interaction') on days when none of the females in the group was in estrus, when one, and when more than one female was in estrus. Also shown in figure 23 are the pooled mean rates of agonistic and nonagonistic social interaction for all adult males combined, the two components of the rate of all social interaction. The 95% confidence intervals for these pooled mean rates also are shown in the figure and the significance of differences in the pooled mean rates when various numbers of females in the group were in estrus was evaluated at the 0.05 level of significance using the variance ratio [COX and LEWIS, 1966].

In figure 23, it will be seen that the pooled mean rate of all social interaction for adult males was significantly lower when one estrous female was present in the group than when none of the females was in estrus. However, the pooled mean rate of all social interaction on days when more than one female was in estrus did not differ significantly from the rate when one female in the group was in estrus. These findings suggest that the rate of social behavior by adult males did not change in relationship to the number of estrous females in the group, but rather to the mere presence or absence of estrous females.

Perhaps more revealing is an analysis of the two components of the rate of all social interaction: the rate of agonistic interaction and the rate of nonagonistic interaction by adult males. Figure 23 shows that the pooled mean rates of agonistic and nonagonistic social interaction for adult males actually changed in opposite directions when estrous females were present in the group. Thus, the pooled mean rate of agonistic interaction (including decided, undecided, and multi-individual agonistic bouts) for adult males was significantly lower on days when one estrous female was present in the group than on days when none of the females in the group was in estrus. When more than one female was in estrus, the rate of agonistic interaction by adult males was still lower, though the increment was not statistically significant. In contrast, the pooled mean rate of nonagonistic interaction for adult males did not differ significantly on days when one estrous female was

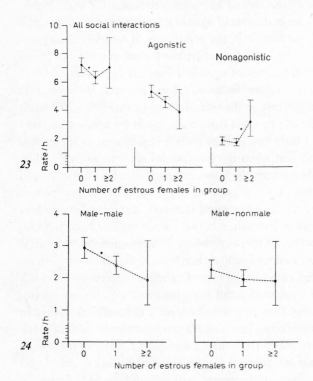

Fig. 23. Rate of social interaction by adult males in relation to number of estrous females in group; showing the pooled mean rates/h of all social interactions, all agonistic interactions, and all nonagonistic interaction for adult males when various numbers of females were simultaneously in estrus as calculated from focal male samples by a method described in the text. Symbols as in figure 14. Raw frequency data from focal male samples can be recovered from this figure by a method described in footnote to table XXXII.

Fig. 24. Rates of agonistic bouts by adult males in relation to number of estrous females in group; showing the pooled mean rates/h of adult male-adult male decided agonistic bouts when various numbers of females were simultaneously in estrus as calculated from focal male sample by a method described in text. Symbols as in figure 14. Raw frequency data can be recovered from this figure by a method described in footnote to table XXXII.

present in the group, compared to days when none of the females in the group was in estrus. However, the pooled mean rate of nonagonistic interaction for adult males was significantly greater when there were at least two estrous females in the group than when only one estrous female was present. Thus, the overall decrease in the rate of all social interaction for adult males that occurred when an estrous female was present in the group was primarily the result of a decrease in the rate of agonistic interaction for adult males.

The observed decrease in the rate of agonistic interaction for adult males on days when at least one female in the group was in estrus is clearly not consistent with the hypothesis that competition for estrous females is accompanied by increased frequency of agonistic interaction in adult males as stated by HALL [1962] for chacma baboons *(Papio ursinus)*. However, it is my impression that the decrease in the rate of agonistic interaction for adult males, in this study, actually resulted from a change in the spacing patterns of adult males when at least one estrous female was present in the group. Thus, it is hypothesized that when estrous females are present in a yellow baboon *(P. cynocephalus)* group, the interadult male distances are increased compared to such distances when none of the females in the group is in estrus, and that the net result of increased intermale distances for adults is a decreased rate of agonistic interaction between adult males. CHANCE [1956] also noted that in his colony of rhesus monkeys *(Macaca mulatta)* intermale distances for adults increased noticeably when females first came into estrus.

If the changes noted above in the rate of agonistic interaction for adult males are the result of changes in adult male behavior with respect specifically to other adult males, then one would expect a concomitant change in the proportion of agonistic bouts in which the bout partner of the focal male was in fact another adult male. Thus, does a decline in the rate of male-male agonistic bouts, in particular, account for the observed decline in the overall rate of male agonistic bouts in the presence of estrous females? Figure 24 is a breakdown of the pooled mean rate of agonistic interaction by adult males into two of its component rates: the rate of decided agonistic bouts between adult males and the rate of decided agonistic bouts with individuals of all other age-sex classes. The rate of agonistic bouts between adult males was significantly lower on days when one female in the group was in estrus than when none of the females was in estrus. When more than one female was in estrus, the pooled mean rate was still lower, though the increment was not statistically significant. In contrast, the pooled mean rate of agonistic bouts between adult males and individuals of all other classes did not change significantly as the number of estrous females in the group increased. Thus, when estrous females were present in the group, the rate of decided agonistic bouts between adult males decreased, but the rate of decided agonistic bouts between adult males and other classes of baboons showed no significant change; in other words, adult male agonistic behavior changed specifically with respect to other adult males, rather than with respect to all other individuals in the group.

Masturbation

Adult males became more likely to masturbate as the number of estrous females in the group increased. Table XXXIII shows that the mean-male rate of masturbation by adult males was greater on days when more than one female in the group was in estrus than on days when only one female was in estrus, and when the pooled mean rates for this behavior were compared the difference proved to be statistically significant (0.05 level, variance ratio [Cox and Lewis, 1966]). Similarly, the rate of masturbation when only one female in the group was in estrus was greater than the rate when none of the females was in estrus, though in this case the difference was not statistically significant.

Wounds

Wounds in Relation to Reproductive States of Females

Wounds are a highly visible and durative index of the intensity of aggression in a social group of baboons. All members of the study group were examined for fresh wounds during the daily census of the group and at other times of the day. Table XXXIV summarizes information on wounding of adult females in various reproductive states as determined from these scan samples as well as the number of female-days of observation for each state. (One female-day of observation accrued for each female present in the group on every day when actual field observations were undertaken.) The proportion of wounds to adult females in the four reproductive states shown – cycling, noncycling, pregnant, and suckling – differed significantly from expectations based on the number of female-days of observation for each state (Chi^2 one sample test, $\chi^2 = 62.9$; d.f. $= 3$, 0.05 level of significance). Cycling females received twice as many wounds than were expected and suckling females received less than one third as many wounds as were expected.

Data on the wounds received by cycling females were examined in

Table XXXIV. Wounds in relation to reproductive state of female, giving the number of wounds received by adult females by reproductive state of female, and associated number of female-days of observation

Reproductive state of female	Female-days of observation (%)		Number of wounds received (%)		Number of wounds expected	Rate, wounds/ female/ day
Cycling	1,233	(34.3)	66	(73.4)	30.9	0.053
Noncycling	221	(6.1)	2	(2.2)	5.5	0.009
Pregnant	760	(21.1)	11	(12.2)	19.0	0.014
Suckling	1,387	(38.5)	11	(12.2)	34.6	0.008
Total	3,601	(100.0)	90	(100.0)	90.0	–
Mean rate						0.025

further detail as follows. The menstrual cycle was divided into five intervals: The first interval of the cycle extended from the day of onset of menstruation through cycle day D-15, and was thus of variable length. The second, third, and fourth intervals of the cycle were from D-14 through D-8, from D-7 through D-1, and from D-day through D + 6, respectively. The fifth interval extended from cycle day D + 7 to the day of onset of menstruation, and was thus also of variable length. If the female became pregnant during a cycle or failed to menstruate, this fifth interval of the cycle was termined on cycle day D + 13. The number of wounds received by cycling females in each of these intervals of the menstrual cycle ist listed in table XXXV as is the number of female-days of observation for each cycle interval.

The proportion of wounds to cycling females in each interval of the menstrual cycle differed significantly from expectations based on the number of female-days of observation for each interval (Chi[2] one-sample test, $\chi^2 = 19.18$; d.f. = 4, 0.05 level of significance). Females in estrus, i.e. cycle days D-7 through D-1, received twice as many wounds than were expected, while deturgescent females, i.e. D-day through D + 13, received almost less than half as many wounds as were predicted. However, if the total of 27 wounds to the swollen sexual skin of cycling females, primarily small cuts and scratches, are excluded from consideration, the proportion of nonsexual skin wounds to females in each menstrual cycle interval did not differ significantly from expectations (Chi[2] one sample test, $\chi^2 = 1.93$; d.f. = 4; 0.05 level of significance). Therefore, although females in estrus are wounded more frequently than expected, the 'extra' wounds are primarily small cuts

Table XXXV. Wounds in relation to cycle state of female, giving the number of wounds received by cycling females by cycle state, and associated number of female-days of observation

Cycle interval	Female-days of observation (%)	Number of wounds received (%)	Number of wounds expected	Rate, wounds/female/day
Menstruation to				
D-15	277 (22.5)	11 (16.7)	14.8	0.040
D-14 to D-18	241 (19.5)	14 (21.2)	12.9	0.058
D-7 to D-1 (estrus)	268 (21.7)	28 (42.4)	14.3	0.104
D-day to D+6	240 (19.5)	5 (7.6)	12.9	0.021
D+7 to menstrua-				
tion or D+13	189 (15.3)	7 (10.6)	10.1	0.037
Unknown	18 (1.5)	1 (1.5)	1.0	0.055
Total	1,233 (100.0)	66 (100.0)	66.0	–
Mean rate				0.053

on the swollen perineum. ZUCKERMAN [1930] states that a plexus of thin-walled blood vessels is situated immediately beneath the epithelium of the sexual skin, and thus some of the small cuts on the sexual skin of swollen females actually may be accidental lacerations, rather than wounds inflicted by other baboons.

Wounds to Adult Males and Nonestrous Females

Table XXXVI lists the number of wounds received by adult males when none of the females in the group was in estrus and when one or more females was in estrus. The number of adult male-days of observation when each number of estrous females was present in the group is given in table XXXVI also. The distribution of wounds to adult males shown in table XXXVI differs significantly from the number expected on the basis of the distribution of adult male-days of observation (Chi^2 one-sample test, $X^2 = 17.69$; d.f. $= 3$; 0.05 level of significance). Adult males were wounded more frequently than expected when one estrous female was present in the group and less than half as frequently as expected when there were no estrous females in the group.

Thus, although figure 24 showed that adult males were less likely to engage in agonistic interaction when estrous females were present in the group, the present analysis of wounding demonstrates that when such

Table XXXVI. Relationship between number of estrous females in group and wounds to adult males, giving the number of wounds received by adult males by number of estrous females in group, and associated number of adult male-days of observation

Number of estrous females	Adult male-days of observation (%)		Number of wounds to adult males (%)		Number of wounds expected	Rate, wounds/ female/ day
0	1,043	(37.8)	7	(14.3)	18.5	0.007
1	1,319	(47.8)	38	(77.6)	23.4	0.029
2	335	(12.2)	3	(6.1)	6.0	0.009
3	61	(2.2)	1	(2.0)	1.1	0.016
Total	2,578	(100.0)	49	(100.0)	49.0	–
Mean rate						0.018

Table XXXVII. Relationship between number of estrous females in group and wounds to adult, anestrous females, giving the number of wounds to adult, anestrous females by number of estrous females in group, and associated number of female-days of observation

Number of estrous females	Anestrous adult female-days of observation (%)		Number of wounds to anestrous females (%)		Number of wounds expected	Rate, wounds/ female/ day
0	1,383	(41.1)	15	(28.8)	21.4	0.011
1	1,521	(45.2)	29	(55.8)	23.5	0.019
2	404	(12.0)	7	(13.5)	6.2	0.017
3	56	(1.7)	1	(1.9)	0.9	0.018
Total	3,364	(100.0)	52	(100.0)	52.0	–
Mean rate						0.015

agonistic interactions did occur, they were more likely to result in the wounding of one of the participants, i.e. to be a serious bout of aggression. The increased likelihood of an adult male-adult male agonistic bout resulting in wounding when an estrous female was present in the group may explain the fact that adult males were also more likely to use nonagonistic behaviors in their interactions with other adult males when estrous females were present in the group (figure 24). In essence it is speculated that an adult male, who might routinely respond to another adult male with agonistic behaviors when none of the females in the group was in estrus, was more likely to respond to another male with a perineal presentation or mounting when estrous females were present in the group. In any event, it is clear that if an adult male did respond to another adult male with agonistic behaviors when estrous females were present in the group, the ensuing agonistic bout was more likely to result in one of the males receiving a wound than if the agonistic bout had occurred when none of the females in the group was in estrus. The greater risk in agonistic bouts when estrous females are present is consistent with the hypothesis (p. 88) that at such times adult males tend to avoid each other.

Table XXXVII lists the number of wounds received by anestrous adult females when various numbers of estrous females were present in the study group. The proportion of wounds to anestrous adult females shown in the table does not differ significantly from expectations based on the number of anestrous adult female-days of observation under each estrous female condition (Chi^2 one sample test, $\chi^2 = 3.32$; d.f. $= 3$; 0.05 level of significance). Thus, while the presence of an estrous female in the group resulted in an increased likelihood of wounding among adult males, it had no detectable effect on the likelihood of wounding of nonestrous adult females.

Summary

1. The dominance rank of estrous females did not change as a result of sexual cycling, nor did consortship with an adult male result in a change in a cycling female's rank. Group changes by adult females were very rare and were not obviously related to the reproductive state of the migrant female.

2. Immigrations of males into the study group and departures of males from the study group were not related to the number of females in estrus in the group at the time. However, periods of inconsistent dominance relationship between adult males occurred less frequently than expected on days when none of the females in the group was in estrus and more frequently than expected when one female in the group was in estrus. Thus, it seems likely that periods of inconsistency in the dominance relationships of adult males, and therefore rank changes by adult males, were related to the presence of estrous females in the study group.

3. The rates of both agonistic and nonagonistic social interaction for adult males were affected by the presence or absence of estrous females in

the study group. Briefly, when estrous females were present in the group the rate of agonistic interaction between adult males decreased. It was hypothesized that when estrous females are present in a yellow baboon group, interadult male distances are greater and thus rates of social interaction lower.

4. Cycling females were wounded twice as frequently as expected on the basis of their numbers in the group, while females suckling infants were wounded less than one third as frequently as expected on the same basis. Among cycling females, females in estrus, i.e. cycle days D-7 through D-1, were wounded twice as frequently as expected and deturgescent females were wounded almost less than half as frequently as expected. Much of the increased wounding of estrous females was small cuts to the swollen sexual skin and may have been accidental lacerations, rather than wounds inflicted by other baboons.

5. Anestrous adult females showed no change in their frequency of wounding as the number of estrous females in the group increased. However, adult males were wounded more frequently than expected when at least one estrous female was present in the group and less than half as frequently as expected when none of the females in the group was in estrus. Thus, although adult male-adult male agonistic interactions were less frequent when estrous females were present in the group, the agonistic interactions that did occur at such times were more likely to result in one of the males receiving a wound than if the agonistic bouts had occurred when none of the females in the group was in estrus.

VI. Conclusions and Speculations

Life History Inferences

In the natural history literature on birds and mammals, the term 'life history' usually refers to a brief description of distribution, breeding habits, and food sources of an animal species [cf. SPRUNT 1955]. However, population biologists reserve for the term life history a much more specific meaning: In a stable population, the proportion of individuals in each age class will be fixed, and the specific distribution of individuals among age classes that is characteristic of a species is referred to as the life history of that species. Additionally, the age-specific birth and death rates for the species, total fecundity, and maximum longevity are also subsumed under the term life history. It is as an extension of this second usage that the term life history is employed in this work.

A particular pattern of life history features is characteristic of each animal species and this pattern has strong consequences for future population growth and development [COLE, 1954]. In turn, the birth rate, death rate, and age structure of a population may be considered consequences of the individual life histories of the members of that population. Thus, the concept of life history relates population biology to the study of individual behavior. The ability of a species to survive in a particular environment or in competition with other species may in fact depend on population phenomena that are a consequence of features of the life histories of individuals; therefore, natural selection would be expected to shape life history patterns as it would any other aspect of the biology of a species [GADGIL and BOSSERT, 1970].

One aspect of the life histories of individuals is their behavior. In fact, the term life history may be extended to include all aspects of an individual's behavior as they affect reproduction and survival. Although this study is concerned primarily with the reproductive success of males of different dominance ranks, it should be emphasized that individuals, not ranks, actually mate and leave offspring. Thus, the results of this study of the relationship between dominance rank and reproductive success must also be discussed in terms of the life histories of individuals.

In chapter II, it was shown that no individual adult male occupied any dominance rank for a very long period of time, at least in comparison to the potential longevity of baboons. In chapter IV, it was shown that each rank had associated with it a specific rate of reproductive success and, furthermore, that the rates of reproductive success for all ranks were not equal. Thus, differential reproduction with respect to dominance rank was demonstrated, though the exact relationship between rank and reproductive success was not adequately predicted by existing models. Also in chapter IV, it was proposed that sexually mature males in a baboon group utilize different reproductive strategies on a short-term basis. The particular strategy that an individual male exhibited at any moment was strongly influenced by his dominance rank though other factors were at work as well. These conclusions will now be discussed as they relate to reproductive success of individuals in a lifetime.

In essence, a male baboon may be viewed as starting his reproductive career confronted with a series of dominance ranks each of which, if occupied, would allow the male to pursue a different short-term reproductive strategy, i.e. has associated with it a specific rate of reproductive success. The challenge that a male faces in his lifetime is to move through these ranks in such a way as to maximize his total reproductive success when compared to other males in the group or population. The process is made somewhat more complex by the fact that males probably do not begin this reproductive competition as equals; the initial dominance rank of a male is probably strongly influenced by the dominance rank of his mother (chapter II). Thus, the particular progression of ranks that a male occupies in his life time may be one of the most important features of his life history, and only through analysis of life histories can one determine if differential reproduction among individuals is taking place in a group or population of animals.

The present study has provided estimates of the rank-specific rates of reproductive success for male baboons in the Amboseli population. However, data on the total lifetime reproductive success for even a single individual nonhuman primate is not presently available. It may be, for example, that every adult male baboon in his lifetime occupies each dominance rank for the same amount of time as does every other male. If so, then, in the long run, all males would be expected to have an equal total lifetime reproductive success. Even if, as is more likely, males differ in the sequence of ranks that they occupy and in the duration of rank occupancy, the total lifetime reproductive success of all males may still be equal. For example, in this study second ranking males had a higher reproductive success than did fifth ranking males, but fifth ranking males had longer durations of rank occupancy, on the average, than did second ranking males. Thus, to achieve any given level of reproductive success, a male may either occupy second rank and reproduce at a high rate for a short period of time or occupy fifth rank and reproduce at a low rate for a longer period of time. The end results in terms of total reproductive success, however, could be equal. Obviously, rank-specific rates of moratilty would play a very important part in determining total reproductive success for these various rank occupancy strategies. Nevertheless, it is clear that different strategies of rank occupancy will not necessarily result in differential reproduction among males.

The development of predictive models of these life history phenomena would greatly aid and stimulate research on differential reproduction and sexual selection. In the case of baboons, male life histories with respect to dominance rank, and thus reproductive success, might be profitably viewed as Markov renewal processes [Cox and LEWIS, 1966]. Duration of rank occupancy data given in figures 9–11, rank change probabilities calculated from table XVII, and rank-specific rates of reproductive success estimated in table XXIV and figures 20 and 21 provide the necessary data to model male life histories as renewal processes. With a model of male life histories, the reproductive consequences of different rank occupancy strategies can be examined and the likelihood of various sequences of rank occupancy determined. Work on such a model is now in progress, though the present research project is still a long way from obtaining a sufficient sample of actual life histories to test any life history model.

Significance of these Results for Other Species

The above life history considerations can be generalized to apply to many biological problems and to a broader range of animals than just baboons. Specifically, in any attempt to demonstrate differential reproduction in a species in relationship to dominance rank or territory possession, one must document the duration of rank or territory occupancy, the reproductive success and mortality rate associated with each rank or territory, the identity of each rank or territory holder, and the identity of nonterritorial or nonreproducing members of the population. Only with these data collected over a substantial portion of the life span of a sample of identifiable individuals can it be demonstrated that differential reproduction is actually occurring within a species. It is important to note that these data are required to demonstrate differential reproduction regardless of any findings of short-term rank or territory differences in reproductive success. These life history considerations also apply to studies that attempt to demonstrate population regulation through behavioral mechanisms [cf. WATSON and MOSS, 1970].

Speculations on the Origin of the Hamadryas Mating System

As noted in chapter I, the genus *Papio* includes two apparently distinct systems of group organization and reproduction. Savannah baboons *(Papio cynocephalus, anubis,* and *ursinus)* typically live and mate in multimale, multi-female groups. In contrast, hamadryas baboons *(P. hamadryas)* have a multilevel organization with most social interaction and reproduction occurring in small one-male units or harems. It seems likely that the organization of hamadryas baboons is part of their adaptation to life in an arid, marginal resource zone. KUMMER [1968], CROOK [1970], CROOK and GARTLAN, [1966], and GOSS-CUSTARD *et al.* [1972] have speculated on the relationship between hamadryas ecology and social organization and their conclusions will not be repeated here.

Briefly, hamadryas society is organized as follows: Large numbers of hamadryas baboons congregate at night on the rock ledges of isolated cliff faces or outcroppings. In the morning, these nocturnal aggregations or 'troops' break up first into multimale 'bands' and then into smaller units composed of one adult male, several females, and the females' offspring. KUMMER [1968] has suggested that the band, or intermediate level of organization, is homologous to the group, or troop, level of organization in savannah baboons. Similarly, aggregations of several groups of savannah baboons have been reported in areas where sleeping sites are restricted [SAAYMAN, 1971 b; ALTMANN and ALTMANN, 1970], and presumably these aggregates are homologous to troops of hamadryas baboons. However, no obvious homologies between the one-male units of hamadryas and organizational features of savannah baboons have yet been proposed. Since most, if not all, reproduction takes place in these one-male units, an understanding of their origin and evolution is fundamental to our understanding of the hamadryas mating system.

Groups of savannah baboons with only one adult male have been reported for each of the savannah baboon species [BOLWIG, 1959; HALL and DEVORE, 1965; HAUSFATER and HAUSFATER, in preparation]. However, analysis of the demography and behavior of one male groups of yellow baboons in the Amboseli Reserve indicates that such groups are probably a chance occurrence representing the tai ends of a shifting group size distribution, rather than a unit in some way homologous to one-male units of hamadryas baboons. Instead it is speculated that hamadryas one-male units are in fact homologous to, and an elaboration of, the temporary male-female consort relationship that is so important for understanding the mating system of savannah baboons (chapter IV). Thus, it is necessary to determine if antecedents of those behaviors considered characteristic of hamadryas one-male units can be found in the male-female consort relationship of savannah baboons.

Although KUMMER [1968] noted that most of the agonistic bouts between adult male leaders of one-male units were caused by disagreements over the possession of a female, it was also noted that agonistic bouts between group leaders were relatively rare. Instead, hamadryas males maintain their harem by the use of threats and overt attacks on their females. If a female strays too far from her unit leader, he will threaten or attack her, and in response she will run toward, and subsequently remain close to, her leader. It was noted in chapter III that Amboseli males herd their female consorts away from competing males, rather than chasing the competing males away from the female. Thus, savannah baboons are already predisposed to a system of organization in which females are maintained by herding them away from other males, rather than though interadult male agonism. Additionally, no two males were ever observed to be simultaneously in consort with a single estrous female in the present study, and thus the savannah baboon consort pair is effectively a temporary one-male unit of organization.

However, as was also noted in chapter III, there are differences in the specific behavior patterns of herding in savannah and hamadryas baboons. Thus, while hamadryas baboon males herd females by biting them on the neck or back, savannah baboon males herd their female consorts by gently nipping or butting them on the flank or hindquarters. Additionally, the response of an Amboseli female to a nip or a butt is to move forward without turning toward the male or in any way decreasing her proximity to him. In contrast, the response of a hamadryas female to a bite or threat from her unit leader is to run directly toward him.

Nevertheless, these differences in herding behavior do not present any major problems in reconstructing the evolution of a hamadryas-type one-male unit from the savannah baboon consort relationship. First, a change in the intensity of agonistic bouts within the consort relationship and concomitant changes in the target area on the body of the female toward which such aggression is directed would produce hamadryas-like behavior on the part of savannah baboon males. Secondly, a characteristic pattern in the agonistic bouts of yellow baboons is the counter-chase (table V B) in which the submissive animal runs toward, rather than away from, his attacker. Thus, all that would be required to produce a hamadryas-like response to herding on the part of savannah baboon females is a change in the context and thresholds for counter-chasing by the female. Finally, all of the consort behaviors on the part of both male and female savannah baboons would have to be maintained even when the female was no longer in estrus to completely arrive at a hamadryas-like one-male unit of organization. Thus, rather than view hamadryas one-male units as a unique or anomalous system of mating and organization within the genus *Papio*, I believe these one-male units are best viewed as an evolutionary specialization, or further elaboration [COUNT, 1958], of a temporary male-female pair bonding, i.e. consortship, that was present in the ancestral baboon population from which hamadryas baboons evolved and which is still seen in savannah baboons today. These speculations concerning the evolutionary relationship of hamadryas social organization to that of savannah baboons are consistent both with the conclusions of JOLLY [1970] based on morphology and natural history and with the conclusions of KUMMER *et al.*, [1970] and NAGEL [1973] based on field experimentation and observation of hybrid behavior.

References

ALLEN, E.; PRATT, J. P.; NEWELL, Q. U., and BLAND, L. J.: Recovery of human ova from the uterine tubes. Time of ovulation in the menstrual cycle. J. amer. med. Ass. *91:* 1018–1020 (1928).

ALTMANN, J.: Observational study of behavior. Sampling methods. Behavior *49:* 227–267 (1974).

ALTMANN, S. A.: A field study of the sociobiology of the rhesus monkey, *Macaca mulatta.* Ann. N.Y. Acad. Sci. *102:* 338–435 (1962).

ALTMANN, S. A.: Primate behavior in review. Science *150:* 1440–1442 (1965).

ALTMANN, S. A.: The pregnancy sign in savannah baboons. Lab. anim. Digest. *6:* 6–10 (1970).

ALTMANN, S. A. and ALTMANN, J.: Baboon ecology. African field research (University of Chicago Press, Chicago 1970).

BALIN, H. and WAN, L. S.: The significance of circadian rhythms in the search for the moment of ovulation in primates. Fertii. Steril. *19:* 228–243 (1968).

BARTHOLOMEW, G. A.: Reproduction and social behavior of the northern elephant seal. Univ. Calif. Publ. Zool. *47:* 369–471 (1952).

BEATTY, R. A.: Fertility of mixed semen from different rabbits. J. Reprod. Fertil. *1:* 52–60 (1960).

BIRDSALL, P. and NASH, D.: Occurrence of successful multiple insemination of females in natural populations of deer mice *(Peromyscus maniculatus).* Evolution *27:* 106–110 (1973).

BOLWIG, N.: A study of the behaviour of the chacma baboon. Behaviour *14:* 136–163 (1959).

BOPP, P.: Zur Abhängigkeit der Inferioritätsreaktion vom Sexualzyklus bei weiblichen Cynocephalen. Rev. suisse Zool. *60:* 441–446 (1953).

BRAWN, U. M.: Aggressive behavior of the cod *(Gadus callarias* L.). Behaviour *18:* 107–117 (1961).

CARPENTER, C. R.: Sexual behavior of free-ranging rhesus monkeys *(Macaca mulatta).* I. Specimens, procedures and behavioral characteristics of estrus. J. comp. Psychol. *33:* 113–142 (1942a).

CARPENTER, C. R.: Sexual behavior of free-ranging rhesus monkeys *(Macaca mulatta).* II. Periodicity of estrus, homosexual, autoerotic and non-conformist behavior. J. comp. Psychol. *33:* 143–162 (1942b).

CARTHY, J. S. and EBLING, F. J.: The natural history of aggression. Institute of Biology Symposia No. 13 (Academic Press, London and New York 1964).

CHALMERS, N. R. and ROWELL, T. E.: Behaviour and female reproductive cycles in a captive group of mangabeys. Folia primat. *14:* 1–14 (1971).

CHANCE, M. R. A.: Social structure of a colony of *Macaca mulatta*. Brit. J. anim. Behav. *4:* 1–13 (1956).

COCKRUM, E. L.: Introduction to mammalogy (Ronald Press, New York 1962).

COLE, L. C.: The population consequences of life-history phenomena. Quart. Rev. Biol. *29:* 103–137 (1954).

COLLIAS, N. E.: Aggressive behavior among vertebrate animals. Physiol. Zool. *17:* 83–123 (1944).

CONAWAY, C. H. and KOFORD, C. B.: Estrous cycles and mating behavior in a free-ranging band of rhesus monkeys. J. Mammal. *45:* 577–588 (1965).

COUNT, E. W.: The biological basis of human sociality. Amer. Anthrop. *60:* 1049–1085 (1958).

COX, D. R. and LEWIS, P. A. W.: The statistical analysis of series of events (J. Wiley & Sons, Chichester 1966).

CROOK, J. H.: The socio-ecology of primates; in CROOK Social behaviour in birds and mammals, pp. 103–166 (Academic Press, London 1970).

CROOK, J. H. and GARTLAN, J. S.: Evolution of primate societies. Nature, Lond. *210:* 1200–1203 (1966).

CUSHING, J. E.: Non-genetic mating preference as a factor in evolution. Condor *43:* 233–236 (1941).

DARWIN, C.: The descent of man and selection in relation to sex (Murray, London 1871).

DARWIN, C.: Sexual selection in relation to monkeys. Nature, Lond. *15:* 18–19 (1876).

DEVORE, I.: Male dominance and mating behavior in baboons; in BEACH Sex and behavior, pp. 266–289 (J. Wiley & Sons, Chichester, 1965).

ECKLAND, B. K.: Evolutionary consequences of differential fertility and assortative mating in man; in DOBZHANSKY, HECHT and STEERE Evolutionary biology, vol. 5, pp. 293–305 (Appleton Century Crofts, New York 1972).

EVANS, L. T.: Courtship behavior and sexual selection of *Anolis*. J. comp. Psychol. *26:* 475–498 (1938).

FISHER, R. A.: The genetic theory of natural selection (Clarendon Press, Oxford 1930).

FREEDMAN, L.: Growth of muzzle length relative to calvorial length in *Papio*. Growth *26:* 117–128 (1963).

GADGIL, M. and BOSSERT, W. H.: Life historical consequences of natural selection. Amer. Naturalist, *104:* 1–24 (1970).

GANDOLFI, G.: Sexual selection in relation to the social status of males in *Poecilia reticulata* (Teleostei: Poeciliidae). Boll. Zool. *38:* 35–48 (1971).

GARTLAN, J. S.: Structure and function in primate society. Folia primat. *8:* 89–120 (1968).

GILLMAN, J.: The cyclical changes in the external genital organs of the baboon *(Papio porcarius)*. S. afr. J. med. Sci. *32:* 342–355 (1935).

GILLMAN, J.: Experimental studies on the menstrual cycle of the baboon *(Papio porcarius)*. VI. The effect of progesterone upon the first parts of the cycle in normal female baboons. Endocrinology *26:* 80–87 (1940a).

GILLMAN, J.: The effect of multiple injections of progesterone on the turgescent perineum of the baboon. Endocrinology *26:* 1072–1077 (1940b).

GILLMAN, J.: Effects on the perineal swelling and on the menstrual cycle of single injections of combinations of estradiol benzoate and progesterone given to baboons in the first part of the cycle. Endocrinology *30:* 54–60 (1942).

GILLMAN, J. and GILBERT, C.: A quantitative study of the nature of the inhibition of oestradiol by testosterone propionate as assessed by the reaction in the perineum of the male and female castrated baboon. S. afr. J. med. Sci. *12:* 87–97 (1942).

GILLMAN, J. and GILBERT, C.: The reproductive cycles of the chacma baboon with special reference to the problems of menstrual irregularities as assessed by the behavior of the sex skin. S. afr. J. med. Sci (Biol. Suppl.) *11:* 1–54 (1946).

GILLMAN, J. and STEIN, H. B.: A quantitative study of the inhibition of estradiol benzonate by progesterone in the baboon *(Papio porcarius)*. Endocrinology *28:* 274–282 (1941).

GOSS-CUSTARD, J. D.; DUNBAR, R. I. M., and ALDRICH-BLAKE, F. P. G.: Survival, mating and rearing strategies in the evolution of primate social structure. Folia primat. *17:* 1–19 (1972).

GRIFFITHS, J. T.: Climate; in MORGAN East Africa. Its people and resources, pp. 107–118 (Oxford University Press, London 1969).

HALL, K. R. L.: The sexual, agonistic, and derived social behaviour patterns of the wild chacma baboon, *Papio ursinus.* Proc. zool. Soc., Lond. *139:* 283–327 (1962).

HALL, K. R. L. and DEVORE, I.: Baboon social behavior; in DEVORE Primate behavior. Field studies of monkeys and apes, pp. 53–110 (Holt, Rinehart & Winston, New York 1965).

HAMMOND, J. and ASDELL, S. A.: The vitality of the spermatozoa in the male and female reproductive tracts. Brit. J. exp. Biol. *4:* 155–185 (1926).

HARTMAN, G. G.: Observations on the viability of the mammalian ovum. Amer. J. Obstet. Gynec. *7:* 40–43 (1924).

Harvard University Computation Laboratory: Table of the Cumulative Binomial Probability Distribution (Harvard University Press, Cambridge 1955).

HAUSFATER, G.: Dominance and reproduction in baboons *(Papio cynocephalus)*. A quantitative analysis; unpublished doctoral dissertation, Chicago (1974).

HAUSFATER, S. A. and HAUSFATER, G.: Demographics and behavior of one-male groups of yellow baboons *(Papio cynocephalus)* (in preparation).

HENDRICKX, A. G.: The menstrual cycle of the baboon as determined by the vaginal smear, vaginal biopsy, and perineal swelling; in VAGTBORG, The baboon in medical research, vol. 2 pp. 437–459 (University of Texas Press, Austin 1967).

HENDRICKX, A. G.: Embryology of the baboon (University of Chicago Press, Chicago 1971).

HENDRICKX, A. G. and KRAEMER, D. G.: Observations on the menstrual cycle, optimal mating time and pre-implantation embryos of the baboon, *Papio anubis* and *Papio cynocephalus.* J. Reprod. Fertil. Suppl. *6:* 119–128 (1969).

HENDRICKX, A. G. and KRAEMER, D. C.: Reproduction; in HENDRICKX Embryology of the baboon, pp. 3–30 (University of Chicago Press, Chicago 1971).

HILL, W. C. O.: Taxonomy of the baboon; in VAGTBORG The baboon in medical research, vol. 2, pp. 3–11 (University of Texas Press, Austin 1967).

HUNTER, R. H. F. and DZIUK, P. J.: Sperm penetration of pig eggs in relation to the timing of ovulation and insemination. J. Reprod. Fertil. *15:* 199–208 (1968).

HUXLEY, J. S.: Darwin's theory of sexual selection and the data subsumed by it, in the light of recent research. Amer. Naturalist *72:* 416–433 (1938).

JOLLY, A.: Hour of birth in primates and man. Folia primat. *18:* 108–121 (1972a).

JOLLY, A.: The evolution of primate behavior (Macmillan, New York 1972b).

JOLLY, C. J.: The large African monkeys as an adaptive array; in NAPIER and NAPIER Old World monkeys. Evolution, systematics and behavior, pp. 139–174 (Academic Press, New York and London 1970).

KAUFMANN, J. H.: A three-year study of mating behavior in a free-ranging band of rhesus monkeys. Ecology *46:* 500–512 (1965).

KAWAI, M.: On the system of social ranks in a natural troop of Japanese monkeys. I. Basic rank and dependent rank; in ALTMANN Japanese monkeys, pp. 66–86 (published by the editor, Atlanta 1965).

KAWAMURA, S.: Matriarchal social ranks in the Minoo-B troop; in ALTMANN Japanese monkeys, pp. 105–112 (published by the editor, Atlanta 1965).

KLOPFER, P. H. and HAILMAN, J. P.: An introduction to animal behavior. Ethology's first century (Prentice Hall, Englewood Cliffs 1967).

KUMMER, H.: Rang-Kriterien bei Mantelpavianen. Der Rang adulter Weibchen in Social-verhalten, der Individualdistanzen und in Schlaf. Rev. suisse Zool. *63:* 288–297 (1956).

KUMMER, H.: Sociales Verhalten einer Mantelpaviangruppe. Beiheft zur Schweiz. Z. Psych. Anwend., No. 33 (Hans Huber, Bern 1957).

KUMMER, H.: Social organization of hamadryas baboons. A field study (University of Chicago Press, Chicago 1968).

KUMMER, H.; GOETZ, W., and ANGST, W.: Cross-species modifications of social behavior in baboons; in NAPIER and NAPIER Old World monkeys. Evolution, systematics, and behavior, pp. 351–363 (Academic Press, New York and London 1970).

LeBOEUF, B. J.: Sexual behavior in the northern elephant seal *Mirounga angustirostris.* Behaviour *41:* 1–26 (1972).

LEVINE, L.; BARSEL, G. E., and DIAKOW, C. A.: Interaction of aggressive and sexual behavior in male mice. Behaviour *25:* 272–280 (1965).

LEWONTIN, R.; KIRK, D., and CROW, J.: Selective mating, assortative mating, and inbreed-ing: definitions and implications. Eugen. Quart. *15:* 141–143 (1968).

LOY, J. D.: Peri-menstrual sexual behavior among rhesus monkeys. Folia primat. *13:* 286–297 (1970).

LOY, J. D.: Estrous behavior of free-ranging rhesus monkeys *(Macaca mulatta).* Primates *12:* 1–31 (1971).

MASLOW, A. H.: The role of dominance in the social behavior of infra-human primates. IV. The determination of hierarchy in pairs and in a group. J. genet. Psychol. *49:* 161–198 (1936).

MICHAEL, R. P.; HERBERT, J., and WELLEGALLA, J.: Ovarian hormones and grooming behaviour in the rhesus monkey under laboratory conditions. J. Endocrin. *36:* 263–279 (1966).

MICHAEL, R. P. and KEVERNE, E. B.: Pheromones in the communication of sexual status in primates. Nature, Lond. *218:* 746–749 (1968).

MICHAEL, R. P. and WELLEGALLA, J.: Ovarian hormones and the sexual behaviour of the female rhesus monkey *(Macaca mulatta)* under laboratory conditions. J. Endo-crin. *41:* 407–420 (1968).

MISSAKIAN, E. A.: Genealogical and cross-genealogical dominance relations in a group of free-ranging rhesus monkeys *(Macaca mulatta)* on Cayo Santiago. Primates *13:* 169–180 (1972).

NAGEL, U.: A comparison of anubis baboons, hamadryas baboons and their hybrids at a species border in Ethiopia. Folia primat. *19:* 104–165 (1973).

NEEL, J. V.; SHAW, M. W., and SCHULL, W. J.: Epidemiology and genetics of chronic diseases (US Department of Health, Education and Welfare, Washington 1965).

NOBLE, G. K.: Sexual selection among fishes. Biol. Rev. *13:* 133–158 (1938).

NOBLE, G. K. and BRADLEY, H. T.: The mating behavior of lizards: its bearing on the theory of sexual selection. Ann. N.Y. Acad. Sci. *35:* 25–100 (1933).

ORIANS, G. H.: On the evolution of mating systems in birds and mammals. Amer. Naturalist *103:* 589–603 (1969).

OVERSTREET, J. W. and ADAMS, C. E.: Mechanisms of selective fertilization in the rabbit. Sperm transport and viability. J. Reprod. Fertil. *26:* 219–231 (1971).

PARKER, G. A.: The reproductive behaviour and the nature of sexual selection in *Scatophaga stercoraria* L. (Diptera: Scatophagidae). IV. Epigamic recognition and competition between males for the possession of females. Behaviour *37:* 113–139 (1970).

RIOCH, D. McK.: Discussion of agonistic behavior; in ALTMANN Social communication among primates, pp. 115–122 (University of Chicago Press, Chicago 1967).

ROSE, R. M.; HOLADAY, J. W., and BERNSTEIN, I. S.: Plasma testosterone, dominance rank, and aggressive behaviour in male rhesus monkeys. Nature, Lond. *231:* 366–368 (1971).

ROWELL, T. E.: Forest living baboons in Uganda. J. Zool., Lond. *149:* 344–364 (1966a).

ROWELL, T. E.: Hierarchy in the organization of a captive baboon group. Anim. Behav. *14:* 430–443 (1966b).

ROWELL, T. E.: Female reproductive cycles and the behaviour of baboons and rhesus macaques; in ALTMANN Social communication among primates, pp. 15–32 (University of Chicago Press, Chicago 1967).

ROWELL, T. E.: Grooming by adult baboons in relation to reproductive cycle. Anim. Behav. *17:* 159–197 (1968).

ROWELL, T.: Baboon menstrual cycles affected by social environment. J. Reprod. Fertil. *21:* 133–141 (1970).

SAAYMAN, G. S.: The menstrual cycle and sexual behaviour in a troop of free-ranging chacma baboons *(Papio ursinus)*. Folia primat. *12:* 81–110 (1970).

SAAYMAN, G. S.: Grooming behaviour in a troop of free-ranging chacma baboons *(Papio ursinus)*. Folia primat. *16:* 161–178 (1971a).

SAAYMAN, G. S.: Behaviour of chacma baboons. Afr. wild Life *25:* 25–29 (1971b).

SAAYMAN, G.S.: Aggressive behaviour in free-ranging chacma baboons *(Papio ursinus)*. J. behav. Sci. *1:* 77–83 (1972).

SADE, D. S.: Determinants of dominance in a group of free-ranging rhesus monkeys; in ALTMANN Social communication among primates, pp. 99–114 (University of Chicago Press, Chicago 1967).

SADE, D. S.: Inhibition of son-mother mating among free-ranging rhesus monkeys. Sci. Psychoanal. *12:* 18–38 (1968).

SADE, D. S.: A longitudinal study of social behavior of rhesus monkeys; in TUTTLE The functional and evolutionary biology of primates, pp. 378–398 (Aldine-Atherton, Chicago 1972).

SCOTT, J. W.: Mating behavior of the sage grouse. Auk *59:* 477–498 (1942).

SKARD, A. G.: Studies in the psychology of needs. Observations and experiments on the sexual need in hens. Acta psychol., Amst. *2:* 172–232 (1937).

SMITH, V. S.; FRICKEN, R.; LATCHAW, P., and GROOVER, M. E.: The influence of the mature male on the menstrual cycle of the female baboon; in VAGTBORG The baboon in medical research, vol. 2, pp. 621–624 (University of Texas Press, Austin 1967).

SNOW, C. C. and VICE, T.: Organ weight allometry and sexual dimorphism in the olive baboon, *Papio anubis;* in VAGTBORG The baboon in medical research, vol. 1, pp. 151–163 (University of Texas Press, Austin 1965).

SPRUNT, A.: North American birds of prey (Bonanza Books, New York 1955).

STOLTZ, L. P. and SAAYMAN, G. S.: Ecology and behaviour of baboons in the Northern Transvaal. Ann. Transvaal Mus. *26:* 99–143 (1970).

SUAREZ, B. and ACKERMAN, D. R.: Social dominance and reproductive behavior in male rhesus monkeys. Amer. J. phys. Anthrop. *35:* 219–222 (1971).

TRIVERS, R. L.: Parental investment and sexual selection; in CAMPBELL Sexual selection and the descent of man 1871–1971, pp. 136–179 (Aldine Publishing, Chicago 1972).

WAGENEN, G. VAN: The coagulating function of the cranial lobe of the prostate gland in the monkey. Anat. Rec. *66:* 411 (1936).

WALLACE, A. R.: Darwinism (MacMillan, London 1889).

WASHBURN, S. L. and DEVORE, I.: Social behavior of baboons and early man; in WASHBURN Social life of early man. Viking Fund Publ. Anthrop. No. 31, pp. 91–105 (Aldine Publishing, Chicago 1961 a).

WASHBURN, S. L. and DEVORE, I.: The social life of baboons. Sci. Amer. *204:* 62–71 (1961 b).

WASHBURN, S. L. and HAMBURG, D. A.: The study of primate behavior; in DEVORE Primate behavior. Field studies of monkeys and apes, pp. 1–13 (Holt, Rinehart & Winston, New York 1965).

WATSON, A. and MOSS, R.: Dominance, spacing behaviour, and aggression in relation to population limitation in vertebrates; in WATSON Animal populations in relation to their food resources. Brit. Ecol. Soc. Symp. No. 10, pp. 167–222 (Blackwell Oxford 1970).

WESTERN, D. and PRAET, C. VAN: Cyclical changes in the habitat and climate of an east African ecosystem. Nature, Lond. *241:* 104–106 (1973).

WILEY, R. H.: Territoriality and non-random mating in sage grouse *Centrocercus urophasianus*. Anim. Behav. Monogr. *6:* 87–169 (1973).

WRIGHT, S.: Systems of mating. III. Assortative mating based on somatic resemblance. Genetics *6:* 144–161 (1921).

ZUCKERMAN, S.: The menstrual cycle of the primates. I. General nature and homology. Proc. zool. Soc., Lond. *1930:* 691–754.

ZUCKERMAN, S.: The menstrual cycle of the primates. III. The alleged breed-season of primates with special reference to the chacma baboon *(Papio porcarius)*. Proc. zoo Soc., Lond *1931:* 325–343.

ZUCKERMAN, S.: The social life of monkeys and apes (Kegan Paul, London 1932).

ZUCKERMAN, S.: The duration and phases of the menstrual cycle in primates. Proc. zool. Soc., Lond., Ser. A *1937:* 315–329.